EVOLUTION OF LAURUSSIA

Evolution of Laurussia

A Study in Late Palaeozoic Plate Tectonics

by

PETER A. ZIEGLER

Geological-Paleontological Institute,
University of Basel, Switzerland
(formerly Shell Internationale Petroleum Maatschappij B.V.,
The Hague, The Netherlands)

Designated publication no. 0163 of the International Lithosphere Programme.
Published with the co-operation of and on behalf of
the Royal Geological and Mining Society of the Netherlands (K.N.G.M.G.)

KLUWER ACADEMIC PUBLISHERS
DORDRECHT / BOSTON / LONDON

Library of Congress Cataloging-in-Publication Data

Ziegler, Peter A.
    Evolution of Laurussia : a study in late Palaeozoic plate
tectonics / by Pieter A. Ziegler.
        p.   cm.
    Includes bibliographical references.
    "Published with the co-operation of the Royal Geological and
Mining Society of the Netherlands (N.N.G.M.G.)."
    "Designated Publication no. 0163 of the International Lithosphere
Programme."

    1. Plate tectonics. 2. Paleogeography--Paleozoic.
3. Paleogeography--Northern Hemisphere.   I. Title.
QE511.4.Z54 1989
551.7'4'091813--dc20   89-24576

ISBN-13: 978-94-010-6696-9   e-ISBN-13: 978-94-009-0469-9
DOI: 10.1007/978-94-009-0469-9

Published by Kluwer Academic Publishers,
P.O. Box 17, 3300 AA Dordrecht, The Netherlands.

Kluwer Academic Publishers incorporates the publishing programmes of
Martinus Nijhoff, Dr W. Junk, D. Reidel, and MTP Press.

Sold and distributed in the U.S.A. and Canada
by Kluwer Academic Publishers,
101 Philip Drive, Norwell, MA 02061, U.S.A.

In all other countries, sold and distributed
by Kluwer Academic Publishers Group,
P.O. Box 322, 3300 AH Dordrecht, The Netherlands.

Printed on acid-free paper

Contents

Preface

This volume aims at providing earth scientists with an overview of the late Palaeozic evolution of North America, Europe, the Arctic and North Africa in a plate tectonic framework.

Plate motions are reviewed that underly the late Silurian-early Devonian suturing of Laurentia-Greenland and Fennosarmatia, the principal constituents of Laurussia, the Devonian and early Carboniferous accretion of Gondwana-derived continental fragments to the southern margin of Laurussia and of Arctica to its northern margin, and ultimately the late Carboniferous to Permian suturing of Laurussia with Gondwana, Kazakhstan and Siberia. An account is given of the evolution of sedimentary basins that developed during late Silurian to late Permian times, within and along the margins of Laurussia. The late Palaeozoic evolution of Laurussia is illustrated by a set of 10 interpretative palaeotectonic-palaeogeographic maps. These are supported by a sequence of reconstructions of the distribution of the continents in the Western Hemisphere.

P.A. Ziegler has taken a multidisciplinary approach to the unraveling of late Palaeozic plate motions and the interaction of continental cratons. At the outset of this study palaeomagnetic, palaeobiogeographical and palaeoclimatological data were collected and reviewed in terms of their compatibility with published palaeoreconstructions of the continents in the Western Hemisphere and the evolution of orogenic belts that evolved through time within and along the margins of Laurussia. Some fundamental discrepancies were recognised between the different data sets that required modifications of the hitherto accepted palaeomagnetically controlled continent reconstructions. As a next step, regional palaeogeographical maps were constructed for North America, Europe and North Africa on the basis of published data and in-house studies of the Shell Group of companies. These maps were digitised and transferred to modified palaeoreconstructions of the continents that were made with the aid of an Evans and Sutherland PS-300 vector graphics terminal.

During the evaluation of the validity of modified palaeoreconstruction of the continents, as presented in this volume, special attention was given to the timing of orogenic events and their relation to intra-plate tectonics, to geodynamic processes governing the evolution of sedimentary basins and to changes in sedimentation patterns and basin outlines. An attempt was made to arrive at a coherent synthesis that is compatible with the principles of plate tectonics, biogeographical and palaeoclimatological data, and, as far as possible, with palaeomagnetic data. The results of this endeavour are summarised in this volume that gives a dynamic account of the evolution of Laurussia from the time of its latest Silurian formation until its late Carboniferous to Permian integration into the Pangea super-continent.

The Shell Group of companies has decided to release this study for publication as a contribution to the advancement of earth sciences in general and to further the efforts of the Inter-Union Commission on the Lithosphere which seeks to elucidate the nature, dynamics, origin and evolution of the lithosphere. P.A. Ziegler is the current chairman of Working Group 3 of the Inter-Union Commission on the Lithosphere; this working group is charged with the study of intra-plate phenomena.

This volume is designated publication No. 0163 of the International Lithosphere Programme.

Den Haag, May 1989
Dr. R.J. Murris, Head of Exploration
Shell Internationale Petroleum Maatschappij B.V.

Acknowledgments

The first part of this volume, discussing the latest Silurian to early Carboniferous evolution of Laurussia was prepared at the invitation of the organising committee of the Second International Conference on the Devonian System. The resulting paper was published in the proceedings of this conference under the title 'Laurussia – the Old Red Continent' (Ziegler, 1988). Simulated by this study, the author could not resist to pursue it further in time till the end of the Palaeozoic. As he arrived at that stage, his retirement from Shell Internationale Petroleum Maatschappij B.V. was imminent and a continuation of the project into the Mesozoic became impractical.

The compilation presented here draws heavily on in-house studies of Shell Internationale Petroleum Maatschappij B.V., Shell Canada Ltd. and Shell Oil Company. Specific contributions to this compilation were made by T. Cook, Shell Oil, on the palaeogeography of the United States, by B.K. Balke, J.H. Braakman and J. de Jager of the joint Shell International and Pecten International Arctic Study Team on the evolution of the Arctic and by E. Lamers, Shell International, on North Africa; these are here gratefully acknowledged. Furthermore, thanks are extended to S. Cameron and C.R. Scotese, Shell Oil, for critical comments on earlier versions of the manuscript and to G. Haug, Shell International for his editorial effort.

Special thanks are extended to Dr. L.P. Zonenshain and Dr. K. Seslavinsky, Moscow and to Dr. A.F. Embry, Calgary, for fruitful discussions on the palaeoreconstruction of the Arctic domain.

The author is indebted to Mrs. Josje Kriest (M.Sc) who took care of the computer data handling and palaeoreconstructions with the aide of an Evans and Sutherland PS300 vector-graphics terminal, to Mrs. Ineke Hilberding for her assistance in data compilation and integration, to Mrs. Wilma Ruggenberg for typing the manuscript and to Messrs. Wim Hoekstra and Ruud van Aarle for drawing the textfigures and plates.

The author acknowledges the magnanimity of Shell Internationale Petroleum Maatschappij B.V. to have given him the opportunity to prepare this paper that was only completed after his retirement from Shell Services, and to have suggested publication of this study. Production of this volume has benefitted from Shell having absorbed the cost of preparing the colour plates.

List of plates

(Following page 102 at end of book.)

Introduction

During the late Caledonian orogenic cycle Laurentia-Greenland and Fennosarmatia (Baltica) became welded together along the Arctic-North Atlantic Caledonian megasuture, thus forming Laurussia, also referred to as the North Continent (Wilson, 1966; Phillips et al., 1976; Roberts and Gale, 1978; A.M. Ziegler et al., 1979; A.M. Ziegler, 1981). Late syn-orogenic and post-orogenic continental clastics, deposited in intramontane, fault-controlled depressions and also in areas peripheral to the Arctic-North Atlantic Caledonides correspond to the Old Red Sand-stone series; these range in age from latest Silurian to late Devonian (Allen et al., 1967; Friend, 1981). From these clastics the landmass occupying the central parts of Laurussia during Devonian times took its name.

The landmass of the Old Red continent, topographically speaking, was fringed to the east by the carbonate shelves of the Barents Shelf and the Moscow Platform, that were associated with a passive margin facing the Sakmarian back-arc ocean (Zonenshain et al., 1984, 1987a and b), and to the west by the carbonate shelves of cratonic North America. The latter faced the Cordilleran miogeoclinal system, that became affected during the latest Devonian to early Carboniferous by the Antler Orogeny (Gilluly, 1963; Gabrielse, 1967; Nilsen and Stewart, 1980). The northern margin of Laurussia was formed by the Inuitian-Lomonosov orogenic system, representing the collision zone between Laurussia and the Arctic Craton (Chukotka-New Siberian Is-

lands). This gradually evolving megasuture became consolidated during the earliest Carboniferous (Trettin and Balkwill, 1979; Kerr, 1981). Throughout Devonian and earliest Carboniferous times the southern margin of Laurussia was occupied by the active, Pacific-type, Appalachian-Variscan geosynclinal system. Its evolution was governed by the northward subduction of oceanic domains separating Laurussia and Gondwana and the accretion of Gondwana-derived continental fragments. Initial contacts between Gondwana and Laurussia were probably established during the latest Devonian. Their super-collision commenced during the late Visean. This marked to onset of the Hercynian Orogeny. The European part of the Hercynian megasuture became consolidated during the Westphalian and its Appalachian-Mauretanian-Ouachita segment during the mid-Permian. During the late Carboniferous and Permian the Kazakhstan and Siberian cratons collided with each other and the eastern margin of Laurussia and became sutured to it along the Ural-Novaya Zemlya-Taimyr fold belt (Ziegler, 1986, 1988).

As a consequence of its progressive integration into the Pangea super-continent Laurussia lost gradually its identity during the late Carboniferous and Permian.

The summary account of the latest Silurian to late Permian evolution of Laurussia, presented in this volume, centres around the discussion of ten palaeogeographic-palaeotectonic maps (Plates 3 to 12). Their topographic bases, giving present day coastlines for stable cratonic areas, have been computer generated. Their projection is orthographic. The age given for each map in millions of years pertains to the continent assembly only (time scale Harland, 1982). The mapping intervals are given in standard stage names.

Continent assemblies shown have been modified from those given by Scotese et al., 1979, 1985 and A.M. Ziegler et al., 1979 and 1981, in order to make them compatible with more recent palaeomagnetic data (Miller and Kent, 1986) and with the tectonic concepts developed during this study (see also Ziegler, 1984, 1986, 1988).

For the ease of reference, a set of the late Silurian to late Permian palaeoreconstructions of the distribution of continents in the Western Hemisphere is given in Plates 1 and 2. These are based on Scotese et al. (1979, 1985), A.M. Ziegler et al. (1979, 1981) Zonenshain et al. (1987a and b) and unpublished maps prepared by Zonenshain, Kuzmin and Natapov. These reconstructions are the bases for the continent distributions shown in the palaeogeographic-palaeotectonic maps shown in Plates 3 to 12.

The palaeoreconstructions given in Plate 1 neglect the palaeomagnetically still poorly constrained Devonian Gondwana apparent polar wander path (APWP) hairpin-loop through

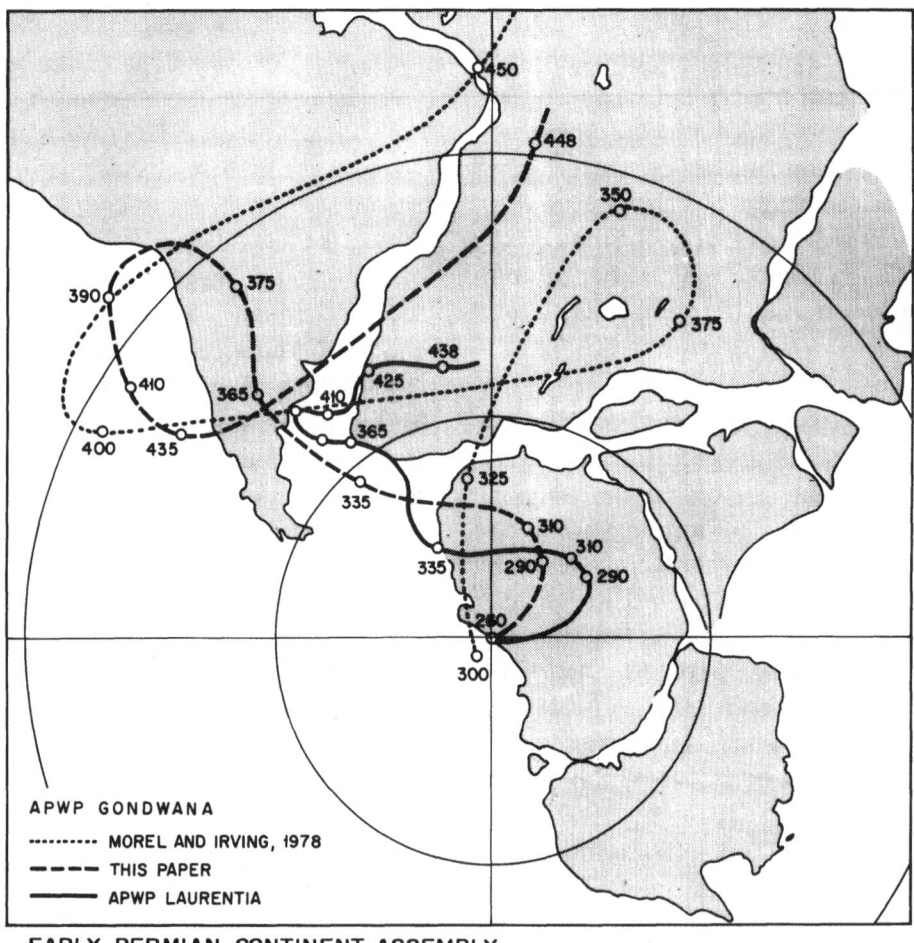

EARLY PERMIAN CONTINENT ASSEMBLY

Fig. 1. Apparent polar wander paths for Gondwana and Laurentia during Silurian to Permian.

Central Africa, as proposed by Morel and Irving's path "Y" (Morel and Irving 1978; Van der Voo, 1987). Instead, a hypothetical APWP is assumed that describes a relatively smooth curve from an early Devonian polar position off the west coast of South America to the generally accepted early Carboniferous pole position near the margin of Antarctica (Fig. 1). This model is compatible with tectonic and stratigraphic evidence, suggesting a latest Devonian-earliest Carboniferous initial collision of Gondwana with Laurussia in the area of Northwest Africa and Iberia (Plate 6).

It is, however, realised that APWP for Laurentia-Greenland, Fennosarmatia, Siberia and Gondwana are still not sufficiently controlled to define relative motions between these cratons to the degree that a reliable and unique model for the evolution of Laurussia can be developed. Palaeolatitudes given in our reconstructions for Laurentia-Greenland are compatible with the palaeomagnetic data given by Dankers (1982) and Miller and Kent (1986). On the other hand the Devonian APWP of Fennosarmatia is not sufficiently constrained to define the magnitude and timing of the proposed Arctic-North Atlantic mega-shear. Nevertheless, an attempt was made to reconcile the geological record of fold belts and sedimentary basins evolving within and along the margins of Laurussia with the available palaeomagnetic and palaeoclimatologic data in order to arrive at an integrated synthesis that is compatible with geodynamic principles. The model presented in this volume, is the result of these endeavours.

The palaeogeographic-palaeotectonic maps given in Plates 3 to 12 are of an interpretative nature, and are even speculative regarding the palinspastic reconstruction of orogenic belts and the facies patterns in basins out of which they developed. For instance, in the Canadian Arctic Archipelago, areas occupied by the Tertiary Eurekan fold belt have been palinspastically restored to fill the gap between Ellesmere Island and the north coast of Greenland that would open when the Baffin Bay-Labrador Sea is closed. Furthermore, as the amount of Mesozoic crustal extension across most of the present-day passive margins is poorly constrained, a relatively loose pre-rifting fit of the continents was chosen. Moreover, there is considerable uncertainty about the original basin outlines and facies developments in areas that were later subjected to broad scale erosion.

In view of this, present day erosional edges of sediments, corresponding to the respective mapping interval, are given with a dotted line in order to make the reader aware of the tenuous nature of the interpretation given. Interpretations presented in these maps have been intentionally pushed beyond the limit of the available data in order to highlight the full implications of the plate tectonic models invoked.

The palaeogeographic information given in Plates 2 to 12 has been abstracted from published literature and in-house studies of the Shell Group of companies. For the USSR these maps are mainly based on Vinogradov (1969) and for cratonic North America on Bassett and Stout (1967), other papers contained in the proceedings of the 1967 International Symposium on the Devonian System (Oswald, 1967), on Ziegler (1969), Cook and Bally (1975), Craig and Waite-Connor (1979), McKee and Crosbey (1975) and unpublished maps of T. Cook (Shell Oil Co.). For the Arctic-North Atlantic domain and Northwest Europe this compilation draws on earlier publications by Ziegler (1982, 1986, 1988) and unpublished maps prepared by the joint Arctic Study team of Bellair Research Center, Houston and Shell Internationale Petroleum Mij. B.V., Den Haag. A chronostratigraphic correlation of Palaeozoic orogenic cycles is given in Figure 2. Often used topographic names are summarised in Plates 13 and 14.

Despite the very generalised and partly hypothetical nature of the maps presented here, and the summary account given in the accompanying text, it is hoped that this volume provides the reader with a rough frame of the latest Silurian to late Permian evolution of Laurussia and its integration into the Pangea super-continent on which more detailed studies can build.

1. Pridolian

The latest Silurian tectonic and palaeo-geographic framework of Laurussia is summarised in Plate 3. Its position relative to Gondwana, and the possible pattern of subduction systems active during this time, is illustrated in Plate 1a.

1.1. Plate Boundaries and Continent Assembly

During the Pridolian-early Gedinnian the Caledonian orogenic cycle terminated with the consolidation of the Arctic-North Atlantic Caledonian fold belt, forming the megasuture between the Laurentia-Greenland and the Fennosarmatian (Baltica) cratons. With this the core of Laurussia was formed (Wilson, 1966; Phillips et al., 1976; Roberts and Gale, 1978; Soper and Hutton, 1984; Roberts, 1988). The continent assembly given in Plate 3, assumes that by Pridolian times Laurentia-Greenland and Fennosarmatia were dextrally offset by some 1000 km as compared to the classical Bullard fit (Bullard et al., 1965).

The late Caledonian consolidation of the Laurussian mega-continent involved the convergence of essentially four plates, namely the Laurentia-Greenland Plate, the Fennosarmatian (Baltica) Plate, the Arctic Plate (Chukotka and New Siberian Islands) and the Proto Tethys-Proto Atlantic Plate. In this setting peripheral plates were the Gondwana Plate, the oceanic Pacific Plate, and the complex Ural Plate of which the Siberian and Kazakhstan cratons formed part (Zonenshain et al., 1987a and b).

As a consequence of the progressive consolidation of the Arctic-North Atlantic Caledonides, and ultimately of the locking of the Iapetus Suture, the Sakmarian-Magnitogorsk arc-trench system developed during the late Silurian within the Ural Ocean. The evolution of this arc-trench system in the southern parts of the Ural Ocean has been reviewed by Zonenshain et al. (1984); its northward continuation into the area of the present-day Kara Sea, as shown in Plate 3, is suggested by Zonenshain et al. (1987b). This new plate boundary separated the oceanic Ural Plate from the Fennosarmatian Plate of which the Sakmarian back-arc ocean now formed part.

The late Silurian tectonic setting of the western margin of the North American craton in unclear. The Cordilleran miogeocline, that was underlain by attenuated continental crust, was flanked to the west by the ill-defined Cassiar-Yukon-Fairbanks platform; this platform was apparently tectonically inactive during the latest Silurian (Gabrielse, 1967; Ziegler, 1969; Morrow and Geldsetzer, 1988). However, an intra-oceanic arc-trench system may have paralleled the western margin of the North American craton (Burchfiel and Davis, 1975).

The northern margin of Laurentia-Greenland and the southern margin of the southward converging continental Arctic Plate (Chukotka and New Siberian Islands) was marked by the gradually evolving Inuitian fold belt. In the

Canadian Arctic Islands this fold belt was probably associated with a south plunging subduction system (Trettin and Balkwill, 1979; Kerr, 1981). It is likely that by late Silurian time the Laurentia-Greenland and Arctic cratons had already collided with each other; however, their mutual collision front presumably did not yet extend into the western Arctic domain.

In our model it was assumed that the boundary between the Arctic Craton and the Lomonosov Ocean, separating it from the northern margin of the Barents Shelf, corresponded to a sinistral oblique subduction zone; the respective arc-trench system is equated with the hypothetical Lomonosov fold belt (Ziegler, 1988).

During the late Silurian, the southern margin of the colliding Laurentia-Greenland and Fennosarmatian plates was formed by a complex arc-trench system that was associated with the north plunging Proto Tethys-Proto Atlantic subduction zone. Convergence of these oceanic domains with the evolving Laurussian mega-continent was accompanied by the northward rafting of a number of continental fragments that became rifted off the northern margin of Gondwana during the Cambro-Ordovician and possibly the early Silurian. Some of these allochthonous terranes, such as the London Platform, the Central Armorican, Saxothuringian and East Silesian cratonic blocks, became accreted to the southern margin of Fennosarmatia already during the Caledonian orogenic cycle. By late Silurian time these continental fragments were enclosed by the North German-Polish, Mid-European and Ligerian-Moldanubian Caledonian fold belts (Ziegler, 1984, 1986). There is evidence that the late Silurian subduction system, paralleling the southern margin of Laurussia, extended from Western Europe eastward into the Black Sea area (Khain, 1984; Samygin and Khain, 1985).

The essentially east-west trending Caledonian fold belts of Western and Central Europe grade westward into the Appalachian geosynclinal system (Bradley, 1983; Williams and Hatcher, 1983; Spariosu and Kent, 1983; van der Pluijm, 1987) and merge in the British Isles and the North Sea with the north-south trending Arctic-North Atlantic Caledonides (Ziegler, 1982; Gayer, 1985; Gee and Sturt, 1986).

By latest Silurian-earliest Devonian time, the composite Traveler-Avalon-Meguma-South Portuguese terrane (Avalonia), the Aquitaine-Cantabrian terrane and the more hypothetical and ill-defined Intra-Alpine, Austro-Alpine and South Alpine-Dinarid terranes were presumably still located some unknown distance to the south of the subduction system marking the southern boundary of Laurussia. These terranes, that are shown schematically in Plates 1 and 3, were separated from Laurussia by the Merrimack-Rheic Ocean and from Gondwana by the Proto Atlantic-Proto Tethys Ocean (Bradley, 1983, Ziegler 1986, Zonenshain et al., 1987b). However, Pridolian and Downtonian faunal similarities between Avalonia and Europe (Boucot et al., 1974) suggest that the Merrimack Ocean was at this time possibly much narrower than shown in Plates 1 and 3. These terranes are all characterised by a Cadomian-Pan African basement complex and a Cambro-Ordovician sedimentary sequence that shows strong lithologic and faunal similarities with Africa. Of importance is particularly the occurrence of late Ordovician glaciomarine deposits in these terranes indicating that they were located by this time in the immediate vicinity of northwest Africa (Powell et al., 1988; Hambrey and Harland, 1981; Fortey, 1984; Cox and Fortey, 1988; Ziegler, 1988).

It is likely that the seafloor spreading system, that was initially responsible for the Ordovician separation of these micro-continents from the northern margin of Gondwana, was still active at the transition from the Silurian to the Devonian and caused the gradual widening of the Proto-Atlantic Ocean at the expense of the closing Merrimack-Rheic Ocean (Bradley, 1983; Keppie, 1985; Kent and Keppie, 1988; Ziegler, 1986, 1987). Schenk (1982) and Bradley (1983) suggest that by late Silurian time a southeast plunging subduction zone had already developed at the leading edge of Avalonia. Indications for contemporaneous transform movements between Avalonia and the Aquitain-Cantabrian terrane are provided by the occurrence of coarse turbidites in Galicia (Martinez-Garcia, 1972) and the shedding of deltaic clastics into the Asturo-Leonese and South Portugese basins (Carls, 1983a).

1.2. Caledonian Fold Belts and Associated Basins

The Arctic-North Atlantic Caledonides extended from the British Isles to northeastern Greenland over a distance of some 4000 km and had a width ranging between 500 and 1000 km. Their structural style is characterised by major basement involving nappes (Roberts and Gale, 1978; Haller, 1971; Hossack, 1985; Hossack et al. 1986; Sturt et al., 1978). In combination with an extensive high-grade metamorphism and a widespread syn- and late-orogenic plutonism (Gee and Sturt, 1986), this indicates that their evolution was accompanied by major crustal shortening, involving crustal delaminations and anatectic remobilisation of lower crustal and partial melting of upper mantle material. The late Ordovician closure of the Iapetus Ocean and the ensuing Himalaya-type collision of Laurentia-Greenland and Fennosarmatia during the late Caledonian diastrophism is thought to have been accompanied by important sinistral translations (Mitchell, 1981; Leggett et al., 1983; Soper and Hutton, 1984).

The existence of a late Caledonian fold belt, transsecting the Barents Shelf is questionable; however, there is evidence for early Caledonian (Ordovician) compressional deformation and magmatism on Severnaya Zemlya and in the northwestern parts of the Taimyr Peninsula. This fold belt , referred to as the Barentsia Caledonides (Plate 3), was apparently inactive during the Silurian, as evident by the sedimentary record of Severnaya Zemlya and Taimyr Peninsula (Gortunov et al., 1984; Khain, 1985; Gramberg et al., 1986).

Geodynamic considerations suggest that the evolution of the Arctic-North Atlantic Caledonides was accompanied by the development of extensive late Silurian foreland basins paralleling their western and eastern deformation fronts (Beaumont et al., 1982; Stockmal et al., 1986). However, large parts of these basins became apparently destroyed during post-Silurian times or are now obscured by younger sediments.

A remnant of the Scandinavian Caledonian foreland basin is still preserved in the Oslo Graben due to latest Carboniferous downfaulting. In it late Silurian marine strata grade upward into Downtonian Old Red molasse series (Holtedahl, 1960; Nilsen, 1973); these became deformed by decollement folding and thrusting during the Gedinnian (Roberts, 1983). On the basis of regional considerations it can be assumed that much of the Fennoscandian Shield was originally covered by early Palaeozoic open marine platform sediments. In areas proximal to the Caledonian thrust front these presumably graded upward into latest Silurian to early Devonian continental red beds accumulating in a foreland basin. Whether parts of such a basin are still preserved in the Barents Sea is unknown.

Similarly, it is likely that the Caledonides of East Greenland, the Hebrides and the Canadian

Maritime Provinces were paralleled by a major foreland basin. In northeastern Greenland and in Ellesmere Island remnants of this foredeep system are preserved. In the partly fault bounded east-west trending Hazen Trough, that lies to the south of the Pearya terrane (forming part of the ancestral Inuitian fold belt), late Silurian clastic turbidites were transported in a westward direction along the basin axis (Hurst et al., 1983; Hurst and Surlyk, 1984; Surlyk and Hurst, 1984; Trettin, 1987). This is indicative of a major clastic source in northeastern Greenland. In turn, this suggests that in the East Greenland foreland basin clastics were transported northward along the basin axis and were deflected into the Hazen Trough. Ice-

cover impedes, however, the recognition of a foreland basin paralleling the deformation front of the East Greenland Caledonides. On trend to the south, a remnant of this foredeep is preserved in Western Newfoundland (Anticosti Basin, Poole et al., 1970). The Anticosti Basin was presumably connected with the northern Appalachian Basin, the sedimentary record of which provides evidence for late Silurian clastics derived from the Appalachian orogen (Cook and Bally, 1975; Dorobek and Read, 1986).

In the northern Appalachian domain, the late Caledonian Salinian disturbance was accompanied by important plutonism (Hatcher, 1985;

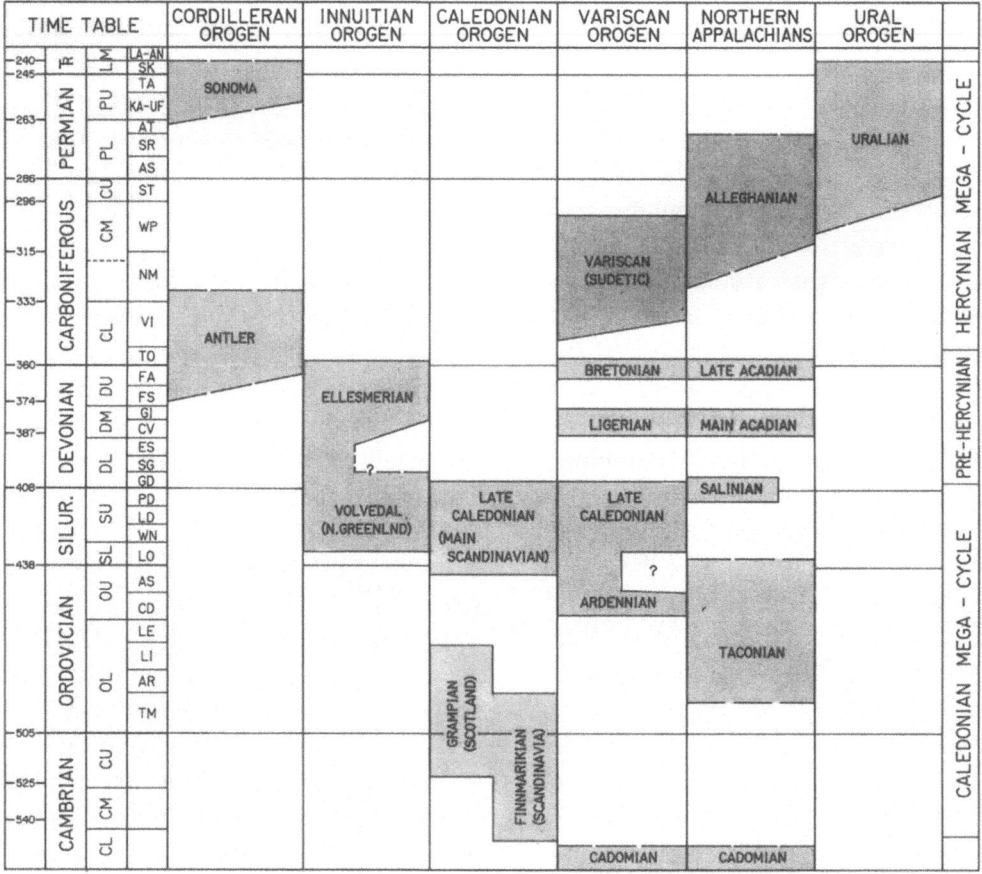

Fig. 2. Correlation of Palaeozoic orogenic cycles.

Hubacher and Lux, 1987). The latest Silurian setting of the New England States and the Canadian Maritime Provinces was characterised by a complex system of arcs and inter-arc basins such as the Greenriver-Sutton anticlinorium, the volcanic Piscataquis arc and the Miramichi Massif facing the Merrimack Ocean (Bradley, 1983). Narrow carbonate shelves flanked intervening deepwater troughs, in which flysch accumulated (Bourque et al., 1986). Furthermore, late Silurian-early Devonian flysch-type series occur in the southern Appalachian Talledega Belt (Tull and Telle, 1988). The northern Appalachian geosynclinal system linked up to the east with the Caledonides of the British Isles and those of continental Europe.

In the intramontane Midland Valley Trough of Scotland, that is closely associated with the Iapetus Suture, deposition of the continental Lower Old Red Sandstone series commenced during the latest Silurian (Bluck, 1978, 1983, 1984; Hutton and Murphy, 1987). On the London Platform, which is underlain by a stable, Gondwana derived cratonic block and which is encircled by Caledonian fold belts, regressive brackish marine clastics conformably overlay marine Silurian strata and grade upward into continental Old Red series (Allen, 1985). The tectonic setting of this basin can be compared to that of a foreland basin that was during Pridolian time still open to a deeper marine trough flanking the Mid-European Caledonides and the Mid-German High to the north (Harz basin). During the Downtonian the London Platform basin became progressively isolated in response to the deformation of the Ardennes foreland basin. Small intramontane latest Silurian basins, containing continental clastics, occur in southwest Ireland and Wales (Gardiner and MacCarthy, 1981).

In the framework of the Caledonian fold belts of Western and Central Europe, the Central Ar-

monican and the composite Saxothuringian-Barrandian Basin, which are also superimposed on Gondwana-derived cratonic blocks, can be considered as 'successor basins'. They are characterised by a nearly continuous marine sedimentary sequence extending from the Cambro-Ordovician into the Devonian and in part even into the early Carboniferous (Guillocheau and Rolet, 1982; Lardeux et al., 1977; Watznauer et al., 1976; Svoboda, 1966). Eastward this complex system of successor basins extends into the Sudetic Basin, southeastern Poland and the Harz area of Germany. These marine basins are located to the east and to the north of the Mid-German High that corresponds to the easternmost parts of the Mid-European Caledonides (Ziegler, 1982, 1984, 1988). During the Pridolian, uplift of the East Silesian and Malopolska massifs, corresponding also to Gondwana-derived continental fragments, was coupled with the subsidence of the Polish-Ukrainian foreland basin, (Tomczyk, 1970; Tomczyk and Tomczykova, 1981; Lewandowski, 1987). The southeastward continuation of this foredeep into the Black Sea area, as shown in Plate 2, is largely conjectural. To the north this foredeep was connected with the carbonate-clastic dominated Baltic Basin, corresponding to the distal parts of the North German-Polish and Scandinavian Caledonian foreland basin (Witkowski, 1979).

1.3. Platform Areas

In cratonic Laurentia-Greenland, North Africa and probably also on the Fennosarmatian platform, Pridolian and Gedinnian times corresponded to a regional regression. As there is no stratigraphic evidence in Gondwana for large-scale late Silurian-early Devonian glaciation (Hambrey and Harland, 1981), despite the circum-polar location of South America (see Plate 1a and Fig. 1), it must be assumed that this regression was of a tectono-eustatic nature (Vail et al., 1977; House, 1983).

During the Ordovician to early late Silurian much of the North American and Greenland cratons were covered by carbonate platforms. In contrast, the latest Silurian Salinian series are preserved as disjointed erosional remnants beneath the regional base-Devonian unconformity (Cook and Bally, 1975). Isolated occurrences of Pridolian carbonates in the Hudson Bay area suggest that during this time marine connections were still maintained between the Arctic Shelf areas and the Michigan-Appalachian carbonate shelf. Devonian subcrop maps indicate that the Trans-Continental and Cincinnati arches and also the Ozark Uplift came into evidence during the latest Silurian and early Devonian (Cook and Bally, 1975). In the Hudson Bay seismic data, calibrated by wells, show that the area was affected by an important phase of block faulting at the transition from the Silurian to the Devonian, causing the upthrusting of a northnorthwest trending high (Dimian et al., 1983; Sanford, 1987; Thorpe, 1988). Similarly there is evidence for contemporaneous faulting in the Moose River Basin in the southern Hudson Bay lowlands (Sanford and Norris, 1973; Telford, 1988). In the Canadian Arctic Archipelago, the Boothia Arch-Cornwallis fold belt, a some 1100 km long north-south trending feature, was affected by a first phase of upthrusting during the Pridolian and Gedinnian (Kerr, 1967, 1977, 1981).

These relatively local tectonic feature, as well as the broad scale lithospheric deflections underlying the uplift of the Trans-Continental Arch, the West Canadian and the Labrador-Greenland highs, indicate that the latest Silurian-earliest Devonian regression was accompanied by important intra-plate deformations; these were presumably induced by compressional stresses that were exerted on the Laurentia-Greenland Craton during the late phases of the Caledonian orogeny.

Pridolian carbonate platforms, flanking the Franklinian deepwater basin (Canadian Arctic Archipelago) and the shelves facing the Cordilleran miogeocline were partly reef fringed. The extent to which the Alaska North Slope Block was covered by a carbonate platform is uncertain. Along the northern Yukon-NW Territories boundary a deeper water shale trough, corresponding to the Richardson Mountains, separated the carbonate platforms of the Mackenzie Shelf and the Peel Plateau (Ziegler, 1969; Morrow and Geldsetzer, 1988).

For the Fennosarmatian platform the palaeogeographic maps of Vinogradov (1969) indicate that its margins were occupied during Ordovician and Silurian times by carbonate platforms, but that much of its central parts were then emergent. The present day distribution of Silurian and Ordovician strata in these shelf basins is limited by the base Devonian unconformity. It is therefore difficult to assess the position of the original Siluro-Ordovician basin edges and the importance of the Pridolian-Gedinnian regression that, similar to Laurentia-Greenland, may also have been accompanied by broad scale lithospheric deflections induced by intra-plate compressional stresses during the terminal phase of the Caledonian orogeny.

On the Barents Shelf, the distribution of latest Silurian carbonates and clastics, as shown in Plate 2, is largely hypothetical as control is limited to Novaya Zemlya, Severnaya Zemlya and the Taimyr Peninsula. Lithofacies distributions shown for Arctica (Chukotka and New Siberian Islands) are highly interpretative (Vinogradov, 1969).

On the Aquitaine-Cantabrian platform, Pridolian series are developed in a generally regressive facies. Similarly a major Pridolian-early Gedinnian regression is evident on the Sahara Platform of Northwest Africa.

2. Early Devonian

The early Devonian megatectonic setting of Laurussia is summarised in the Emsian palaeotectonic-palaeogeographic map given in Plate 4. The position of Laurussia relative to Gondwana is illustrated by the palaeoreconstruction of continents in the Western Hemisphere shown in Plate 1b.

2.1. Plate Boundaries and Continent Assembly

The early Devonian evolution of Laurussia reflects a major reorganisation of plate boundaries. With the Gedinnian locking of the Iapetus suture and the consolidation of the Arctic-North Atlantic Caledonides, this long-standing plate boundary between Laurentia-Greenland and Fennosarmatia became inactive. At the same time the importance of the Sakmarian-Magnitogorsk arc-trench system, as the eastern boundary of the Laurussian Plate, became emphasised (Zonenshain et al., 1987b). Although the Pacific margin of the North American craton, corresponding to the outer plateau of the Cordilleran miogeocline, was apparently tectonically inactive during the early Devonian (Gabrielse, 1967), there is evidence that it was paralleled by an active intra-oceanic arc system, referred to as the Klamath-Sierran arc (Burchfiel and Davis, 1975).

During the early Devonian continued northwestward subduction of the Merrimack-Rheic Ocean along the Appalachian-Ligerian-Moldanubian arc-trench system was ac-

companied by the convergence of the Traveler-Avalon-Meguma-South Portuguese (Avalonia), the Aquitaine-Cantabrian and the Intra-Alpine terranes with the southern margin of Laurussia. By Emsian times the width of the oceanic basins separating these micro-continents from Laurussia was drastically narrowed and their full scale collision with the latter was imminent. This is supported by phyto-geographic links between Laurussia and Avalonia (Raymond, 1987). The subduction zone at the leading edge of Avalonia continued to be active as evident by a tectonothermal event ranging in age from 425 to 390 Ma (Bradley, 1983; Dallmeyer and Keppie, 1987). Similar subduction systems may have become active along the northern margin of the Aquitaine-Cantabrian and Intra-Alpine terranes. Transform movements between Avalonia and the Aquitain-Cantabrian terranes persisted as evident by the occurrence of coarse turbidites in Galicia (Martinez-Garcia, 1972) and the shedding of deltaic clastics into the Asturo-Leonese Basin (Carls, 1983a) from the rising Central Iberian ranges.

Apparent widening of the Proto-Atlantic Ocean, as shown on Plate 4, could be related to continued activity along the seafloor spreading system that was responsible for the separation of the above terranes from Gondwana. This supposition is, however, highly dependent on the assumed APWP of Laurussia and Gondwana (Fig. 1).

During the early Devonian, Laurentia-Greenland remained in an equatorial position (see Plate 1) whereas sinistral movements along the Arctic-North Atlantic fracture zone accounted for a northward translation of Fennosarmatia. In our model it was arbitrarily assumed that these movements amounted to about 200 km, during the early Devonian so that by Emsian time the dextral offset between Greenland and Norway had decreased to about 800 km.

Movements along the Arctic-North Atlantic mega-shear were paralleled by an important phase of back-arc rifting in Western and Central Europe where the Rhenohercynian Basin subsided rapidly in the area of the Mid-European Caledonides. Back-arc extension proceeded apparently to crustal separation and the opening of oceanic basins (Ziegler, 1982, 1984, 1988). Development of a back-arc extensional system in this area can be related to a decrease in the convergence rate between the Rheic Ocean and the southern margin of Fennosarmatia, thus permitting the ascertion of a back-arc convection cells (Uyeda, 1981). In this respect, the contemporaneity of the sinsitral translation between Laurentia-Greenland and Fennosarmatia and of the opening of the at least partly oceanic Rhenohercynian back-arc basin suggests a possible genetic link between the two processes.

Evidence for early Devonian opening of a back arc-ocean in the Rhenohercynian Basin comes from southern England (Lizzard complex, Badham, 1982; Isaac and Barnes, 1985) and also from the Rheinische Schiefergebirge in Germany (Engel et al., 1983; Platen et al., 1989; Franke et al., 1989). Moreover, circumstantial evidence is provided by the occurrence of early Carboniferous I-type granites in the Vosges and Odenwald. These are thought to be related to a south plunging subduction zone which was active during the early Carboniferous closure of the Rhenohercynian Basin (Holl and Altherr, 1987; Volker and Altherr, 1987; Henes-Klaiber

et al., 1989).

Opening of the Rhenohercynian back-arc basin was accompanied by further compressional deformations in southern Poland, as evident by the shedding of Siegenian and Emsian molasse series from the Malopolska Massif into the continuously subsiding late Caledonian foreland basin of southeast Poland and the Ukraine (Ziegler, 1982, 1988; Hizhnyakov and Pomyanovskaya, 1967). These deformations are probably the effect of differential movements between the Central Armorican-Saxo-thuringian-East Silesian Block relative to the stable margin of the Fennosarmatian platform.

Early Devonian northward translation of Fennosarmatia relative to Laurentia-Greenland implies the gradual narrowing of the hypothetical Lomonosov Ocean that is thought to have separated the northern margin of the Barents Shelf from the equally hypothetical Lomonosov fold belt marking the eastern margin of the Arctic Craton. Continued early Devonian convergence of Arctica with the northern margin of Laurussia, presumably in response to continued sea-floor spreading in the Proto-Arctic Ocean (Zonenshain et al., 1987b), caused the westward propagation of their mutual collision front. Sinistral movements along the Arctic-North Atlantic mega-shear were apparently taken up within the evolving Inuitian fold belt.

2.2. Basins Associated with Arctic-North Atlantic Mega-shear

In the domain of the Arctic-North Atlantic Caledonides and that of the British Isles, Old Red Sandstone series accumulated during the early Devonian in often widely separated, fault-bounded, rapidly subsiding intra-montane basins (Friend, 1981). Main depocentres in the British Isles were the Munster Basin -Dublin -

Northumberland Trough and the Midland Valley Graben. Both extend from Ireland to the coast of the North Sea. A further important early Devonian basin is the large, complex Orcadian Basin on the Shetland Shelf (House et al., 1977, Leeder, 1976, Bluck, 1978, 1984; Coward and Enfield, 1987). The latter probably extends across the North Sea into the Hornelen and Solund basins of the coastal area of southern Norway (Nilsen, 1973; Steel, 1976; Steel and Gloppen, 1980; Ziegler, 1982; Torsvik et al., 1987). Seismic data from the West Shetland Shelf indicate the presence of a complex set of fault bounded Old Red basins, that presumably came also into evidence during the early Devonian (Enfield and Coward, 1987).

The development of these largely tensional basins, in response to post-orogenic gravitational collapse of the Caledonides, combined with diverging wrench movements, was accompanied by extensive intrusive and extrusive igneous activity. This indicates that their evolution was accompanied by deep crustal fracturing. Lower Old Red clastics reach a maximum thickness of 6600 m in the Midland Valley Graben.

In the coastal area of Mid-Norway (Trondheims Fjord) early to middle Devonian continental conglomeratic series are preserved in a narrow trough (Nilsen, 1983; Roberts, 1983; Norton, 1986). In the adjacent offshore area reflection seismic data indicate the presence of thick pre-Permian series, contained in a half graben, that may contain Devonian and early Carboniferous clastics (Bukovics and Ziegler, 1985).

In Central East Greenland up to 7000 m thick Emsian to early Tournaisian Old Red series accumulated in the Trail Ø-Hudson Land area (Haller, 1971; Friend et al., 1976). This basin probably plunges southward under Permo-Car-boniferous and Mesozoic strata exposed on Jameson Land (Ziegler, 1988).

In the Central Spitsbergen Trough 1500-2000 m thick Downtonian red beds, deposited on a late Caledonian basement complex, became folded and thrusted during the early Gedinnian Haakonian deformation phase. Sedimentation resumed during the late Gedinnian with continental red beds grading upward into marginally marine middle Devonian clastics. These strata, that attain a thickness of some 6000 m, were shed from south-western and south-eastern sources (Birkenmajer, 1981; Steel and Worsley, 1984). Similar series may be present in fault bounded troughs beneath the Permo-Carboniferous and Mesozoic cover of eastern Svalbard, the Svalbard Bank and possibly also in the western Barents Sea (Rønnevik and Beskow, 1983; Rønnevik and Jacobsen, 1984).

Outcrop and reflection seismic data indicate that the Lower Old Red Sandstone basins of the Arctic-North Atlantic area subsided under a tectonically active, diverging wrench regime. This suggests that post-orogenic tensional collapse of the Caledonian fold belt was accompanied by regional uplift, its rapid erosional unroofing and movements along the postulated Arctic-North Atlantic mega-shear.

2.3. Hercynian Geosynclinal System

The Hercynian geosynclinal system was associated with the northwestward plunging Merrimack-Rheic subduction zones that marked the southern margin of Laurussia; it consisted of the Appalachian arc-trench system and the complex arc-trench and back-arc basin system out of which the Variscan chains of Europe evolved during the late Carboniferous.

The early Devonian development of the Variscan geosynclinal system was governed by a major phase of back-arc extension that caused the collapse of the Mid-European Caledonides and the subsidence of the Rhenohercynian Basin. This newly developing basin was separated from the Central Armorican-Saxothuringian successor basins to its south by the Normannian-Mid-German High. In their eastern parts the Rhenohercynian and the Saxothuringian basins were in open communication with each other. The Central Armorican Basin was probably also open to the west to the Rheic-Merrimack Ocean. Similarly the western parts of the differentially subsiding Rhenohercynian Basin were in communication with this ocean, presumably across the Normannian High (Ziegler, 1982, 1988).

Early Devonian subsidence of the Rhenohercynian basin was accompanied by tensional tectonics and an alkaline, basaltic volcanism. First marine incursions, originating from the west and east, are evident during the late Gedinnian and by Siegenian times deep-water conditions were established. Crustal separation and the onset of sea-floor spreading occurred presumably during this time (Badham, 1982; Isaac and Barnes, 1985; Rolet et al., 1986; Engel et al., 1983; Platen et al., 1989; Franke et al., 1989).

Only the eastern parts of the Rhenohercynian Basin are characterised by continuous deep-water sedimentation across the Silurian-Devonian boundary. Elsewhere basal transgressive clastics overlay unconformably Silurian and older strata or Caledonian metamorphics and intrusives. Throughout early Devonian times the northern margin of the Rhenohercynian Basin was occupied by major deltaic complexes,that were sourced by rivers originating on the Old Red landmass. This clastic fringe extended from Ireland to southern Poland and possibly to the Black Sea area. Basinal areas were characterised by black shales, indicating sediment starvation commencing generally during the Siegenian. Clastic supply to the basin from the Normannian-Mid German High abated about at the same time (Meyer and Stets, 1980; Walliser and Michels, 1983; Ziegler, 1982, 1988).

In the Central Armorican Basin, early Devonian shallow marine carbonate and clastic series conformably overlay late Silurian shales, whilst continuous deepwater sedimentation characterised the Saxothuringian Basin. In both basins there is evidence for early Devonian rift-related volcanism, suggesting that also these areas were affected by tensional stresses. However, back-arc extension in these basins was apparently less intense than in the Rhenohercynian Basin and presumably did not proceed to crustal separation (Rolet et al., 1986; for summary stratigraphic columns, see Ziegler, 1988).

For the southeast European parts of the Variscan geosynclinal system only limited information is available due to their severe overprinting during the Variscan and Alpine orogenies. For the Caucasus area Adamia et al. (1981) and Samygin and Khain (1985) visualise a complex arc-trench system.

In the Appalachian geosynclinal system there is, unlike in the Variscan geosyncline, no evidence for major early Devonian back-arc extension. Clastic influx into the Appalachian Basin from the nascent orogen was at a generally low level during the Gedinnian, increased during the Siegenian (Oriskany Sandstone) and abated altogether during the Emsian-early Eifelian. By this time platform areas were occupied by carbonate shelves giving way basinward to a sediment-starved trough (Oliver et al., 1967; Dorobek and Read, 1986). Evidence for back-arc extension at the transition from the early to the middle Devonian comes only from the southern Appalachian Talladega basin; in this basin the accumulation of coarse turbidite fans was accompanied by the extrusion of bimodal volcanics (Tull and Telle, 1988). In contrast, in the northern Appalachian Gaspé Trough, syn-orogenic turbiditic series graded upward into Siegenian deltaic and continental sands (Rust, 1981) whilst in the Merrimack Trough, flanking the volcanic Piscataquis Arc to the east, the accumulation of deeper water clastics persisted. This illustrates that despite the incipient closure of the Merrimack Ocean the Appalachian arc system apparently did not become substantially and progressively accentuated. On the other hand, the Emsian decrease in clastic influx into the Appalachian foreland basin could also be interpreted as having been caused by the deepening of its proximal, orogen-ward, parts in response to tectonic loading during precursor phases of the Acadian Orogeny (Bradley, 1983).

2.4. Inuitian Orogen and Franklinian Basin

During the early Devonian continued convergence of Arctica and Laurussia can be inferred from increased clastic influx from northern sources into the Franklinian deepwater basin, corresponding to the foredeep of the rising Inuitian fold belt. By Emsian time turbidite fans had propagated into the area of Ellef Ringess and Mackenzie King islands and the basin began to shallow out (A. Embry, personal communication). Additional evidence for orogenic activity comes from northern Ellesmere Island where the Cape Woods granite has been dated as 390 ± 10 Ma (Trettin et al., 1987) and where early Devonian shallow marine clastics form part of a neo-autochthonous sequence (U. Mayr, personal communication). Furthermore, on the southern carbonate shelf of the Franklinian Basin, the Boothia Arch-Cornwallis fold belt became accentuated during the early Devonian (Kerr, 1977, 1981), and in southeastern Ellesmere Island the Inglefield

Arch became uplifted during the Emsian (Smith and Okulitch, 1987). These deformations can be related to intra-plate compressional stresses that were exerted onto the foreland during the early Devonian orogenic phases of the Inuitian fold belt. Concomitant southwestward thrusting of the Pearya Terrane was accompanied by sinistral strike slip motions (Trettin, 1987), compensating for similar movements along the Arctic-North Atlantic mega-shear.

In the northern foreland of the developing Inuitian fold, belt the occurrence of early Devonian clastics on Wrangel Island may be taken as evidence for the proximity of a rising fold belt. Palaeogeographically speaking, these clastics are offset to the north on Chukotka by shallow marine sands and shales; these may also have been derived from the Inuitian orogen (Vinogradov, 1969). Furthermore, possible Devonian clastics have been reported from Henrietta Island, located to the northeast of the New Siberian Islands, whereas on the latter early Devonian carbonates conformably overlay late Silurian carbonates (Fujita and Cook, 1986).

The westward continuation of the Inuitian fold belt and its connection to the Cassiar-Yukon-Fairbank Platform, as show in Plate 4, is uncertain (Gabrielse, 1967). Similarly there is no tangible evidence in support of the Lomonosov fold belt marking the eastern margin of the Arctic Craton. Areas corresponding to this fold belt form now the submarine Lomonosov Ridge.

2.5. Platform Areas

On the North American Craton, and also in North Africa, the Pridolian-Gedinnian regression was followed by cyclical transgressions that inundated by Emsian times already large areas. On the Fennosarmatian Platform there is no evidence for contemporaneous major transgressions. The sparse data available from the Arctic Craton do not permit an assessment of the importance of the earliest Devonian regression and the Siegenian-Emsian transgression.

Earliest Devonian times are thought to correspond to a period of global lowstand in sea-level. The late Siegenian and Emsian transgression, particularly evident in North American, may correspond to a precursor of the general middle Devonian rise in sea-level, but may in part also be related to the relaxation of the intra-plate compressional stresses that had induced lithospheric deflections during the late Caledonian diastrophism and to the decay of lithospheric thermal anomalies (Vail et al., 1977; House, 1983; Johnson et al., 1985).

The early Devonian collapse and erosional unroofing of the Arctic-North Atlantic Caledonides was apparently coupled with an uplift of their foreland areas. Through time, this induced the progressive destruction of the foreland basins which are thought to have paralleled this fold belt during the late Silurian and early Gedinnian. At the same time, the formerly extensive Siluro-Ordovician sedimentary cover of the Labrador-Greenland High (Cook and Bally, 1975) and presumably also of the Fennosarmatian High, became truncated and ultimately completely removed. Recent fission track analyses from southern Sweden indicate, however, that the main phase of its uplift and the removal of supra-crustal series occurred during the Triassic (Zeck et al., 1988); whether this applies on regional scale is unknown.

Other areas of extensive early Devonian erosion are the Trans-Continental Arch and the West Canadian High (Cook and Bally, 1975; Bassett and Stout, 1967). To judge by the areal distribution of the early Devonian erosional/non-depositional hiatus, corresponding to the Tippecanoe-Kaskaskia sequence boundary of Sloss (1963), the Old Red landmass had reached its maximum extent during the Gedinnian and Siegenian. This huge continent straddled the equator and extended in an east-west direction over some 7500 km and in a north-south direction over 3000 to 4000 km. The north-south trending Caledonian mountain system, forming a continental divide, had presumably a considerable effect on the aridity, particularly of the western parts of Laurussia. The high rate of clastic influx into the Rhenohercynian Basin suggests that an extensive drainage system had developed on the southern slopes of the British and North Sea Caledonides. Overall, the Old Red landmass was probably characterised by a hot, seasonally arid climate.

During the earliest Devonian marine sedimentation was generally restricted to areas corresponding to the latest Silurian outer shelves (e.g. Ural, Franklinian, Cordilleran and Appalachian shelves) and to the deeper parts of differentially subsiding intra-cratonic basins (e.g. Illinois Basin).

On the North American Craton, late early Devonian transgressions began to encroach onto the Laurentian landmass from the Franklinian Basin in the north and from the Appalachian geosynclinal system in the south and south-east. During the Siegenian and Emsian, cyclical transgressions began to inundate the area of the West Canadian High, which presumably started to subside differentially. By late Emsian time the Bear Rock-Lower Elk Point Basin occupied large parts of Western Canada, thus forming the Canadian Shelf. Its extent, as shown of Plate 4 is hypothetical. Erosional remnants of this shelf basin are preserved in the subsurface of the Northwest Territories, northeastern British Columbia, Alberta and Saskatchewan (Bassett and Stout,

1967; Ziegler, 1969) and in the central Hudson Bay (Dimian et al, 1983). In the latter the onlap-relationship of early Devonian carbonate series suggests that the late Caledonian fault systems had become inactive. The distal parts of the Canadian Shelf, facing the Franklinian and Cordilleran deepwater basins, were occupied by carbonate platforms, whilst in its interior parts restricted carbonates and evaporites were deposited. To the southwest this basin was limited by the Peace River-Alberta-Central Montana Arch and to the south by the Trans-Continental Arch, which presumably linked up with the Labrador-Greenland High. In Saskatchewan the southward advance of transgressions was limited by the topography of the erosional Meadow Lake Escarpment beyond which the thin Ashern red beds may have already started to accumulate on the Prairie Plateau during the late Emsian.

During Emsian transgressions originating from the Appalachian Shelf invaded the Grate Lakes area (Michigan Basin) and may have advanced as far west as Iowa. The by now expanded carbonate shelves probably extended northward into Ontario and Quebec. In the Moose River Basin (southwestern Hudson Bay), the Kenogami River red beds and the overlaying Sextant clastics and Stooping River carbonates may represent a basin margin facies (Sanford and Norris, 1973; Telford, 1988). On the other hand, the area of the Trans-Continental Arch and the Ozark Uplift remained positive features throughout early Devonian times and were the locus of erosion and karstification of early Palaeozic carbonates (Sandberg and Maple, 1967).

On the Fennosarmatian Platform the record of early Devonian sedimentation is restricted to the Baltic area, where thin lacustrine-brackish red beds were deposited in a shallow depression. The remainder of this large landmass corresponded probably to a low relief peneplain from which only limited clastics were shed onto the adjacent, generally narrow shelves. It is uncertain whether the Donets Graben, that could be considered as forming part of the Variscan back-arc extensional system, had already started to subside differentially during the early Devonian (Vinogradov, 1969).

Facies patterns shown for the Barents Shelf are conjectural; however, a wide carbonate shelf, containing an evaporitic basin in the area of Severnaya Zemlya, occupied its eastern and possibly also its northern margins (Vinogradov, 1969) facing the Sakmarian and the hypothetical Lomonosov oceans. Based on limited control it is assumed that by late early Devonian times the northern parts of the Arctic Craton were occupied by carbonate shelves facing the Proto-Arctic Ocean.

3. Middle Devonian

The middle Devonian setting of Laurussia is summarised by the Givetian palaeotectonic-palaeogeographic map given in Plate 5 and its possible position relative to Gondwana is depicted by the palaeoreconstruction of continents in the Western hemisphere shown in Plate 1c.

Our model for the early and middle Devonian relative motion between Laurussia and Gondwana differs fundamentally from models honouring the still ill-defined middle and late Devonian Gondwana APWP hair-pin loop through Central Africa (Fig. 1; Morel and Irving, 1978; Hargraves et al., 1987). These models imply that Gondwana collided with Laurussia during the early Devonian and receded rapidly from it during the middle and late Devonian, thus causing the re-opening of the Proto Atlantic and Proto Tethys oceans (Van der Voo, 1987). As this concept is difficult to reconcile with the middle and late Devonian evolution of the Appalachian-Variscan geosynclinal system and of the Sahara Platform, these palaeomagnetic data have not been honoured in the model presented here. Obviously, more control points are needed to underpin the Gondwana APWP hair-pin loop through Central Africa before it can be accepted at face value.

3.1. Plate Boundaries and Continent Assembly

The evolution of Laurussia during the middle Devonian continued essentially along lines established during the early Devonian. Important changes in plate boundaries occurred, however, in the area of the Hercynian geosynclinal system.

Late Emsian to Eifelian complete closure of the Merrimack Ocean and the ensuing full scale collision of the Traveler-Avalon-Meguma-South Portuguese composite terrane with the Appalachian arc-trench system gave rise to the Acadian Orogeny during which these continental blocks were accreted to the southern margin of Laurussia (Poole, 1977; Schenk, 1978, 1982; Bradley, 1983; Williams and Hatcher, 1983; Baker and Gayer, 1985; Keppie, 1985). At about the same time the Rheic Ocean became closed and the Aquitaine-Cantabrian and the more hypothetical Intra-Alpine terranes collided with the arc-trench system paralleling the southern margin of the Ligerian-Moldanubian Cordillera, to which they became sutured during the Ligerian Orogeny (Bernard-Griffiths et al, 1977; Autran and Cogné, 1980; Autrand and Dercourt, 1980; Matte, 1983, 1986). These two docking events are broadly synchronous (Fig. 2; Ziegler, 1984, 1988). Furthermore, during the Acadian-Ligerian Orogeny the Aquitaine-Cantabrian and the South Portuguese micro-cratons became sutured along the Central Iberian fold belt (Martinez-Garcia, 1980; Iglesias et al., 1983; Chacon et al., 1983). Whether contemporaneous deformations occurred also along the southeastern margin of Fennosarmatia is uncertain.

Docking and accretion of these Gondwana

derived terranes to the southern margin of Laurussia was followed, probably already during the late middle Devonian, by the development of a new northwest plunging subduction zone, paralleling the southeastern margin of Avalonia, along which the lithosphere of the now gradually closing Proto-Atlantic Ocean became consumed. Whether at the same time similar subduction zones developed also along the southern margins of the Aquitaine-Cantabrian and intra-Alpine terranes is uncertain.

During the Acardian-Ligerian Orogeny, back-arc extension in the Variscan geosynclinal system was temporarily interrupted. Moreover, our model assumes that movement along the Arctic-North Atlantic mega-shear accelerated during the middle Devonian to the extent that by late Givetian time the dextral offset between Greenland and Norway was reduced to some 400 km. This implies progressive closure of the Lomonosov Ocean and the onset of collision between the Barents Shelf and the southeastern margin of the Arctic Craton. Sinistral translations between the Laurentia-Greenland and the Fennosarmatian sub-plates were probably taken up in the Inuitian fold belt, in which orogenic activity persisted due to continued convergence of Arctica with Laurussia.

At the eastern margin of Laurussian Plate, intensified late Emsian to Eifelian subduction processes along the Sakmarian-Magnitogorsk arc were paralleled by a phase of back-arc compression, causing submarine obduction of oceanic crust onto the lower slope of the Fennosarmatian margin. Back-arc compression ceased, however, during the late Eifelian and gave way to back-arc extension. During the Givetian, the formation of new oceanic crust in the Sakmarian back-arc basin was paralleled by intra-continental rifting on the eastern parts of the Fennosarmatian platform (Ruzencev and Samygin, 1979; Artyushkov and Baer, 1983;

Zonenshain et al, 1984; L.P. Zonenshain, personal communication).

The western margin of the North American craton became tectonically active during the late middle Devonian as evident by the shedding of turbidites from the outer platform into the Cordilleran miogeocline (Gabrielse, 1967; Morrow and Geldsetzer, 1988; Gordey, 1988). This is thought to reflect accelerated convergence of the Pacific Plate with the North American Craton, causing increased activity along the intra-oceanic Klamath-Sierran arc-trench system and compressional deformation of the associated back-arc basin (Burchfiel and Davis, 1975).

3.2. Arctic-North Atlantic Mega-shear and Associated Basins

The stratigraphic record of the Anticosti Basin suggests that in the Gulf of St. Lawrence the Caledonian thrust front became re-activated during the Acadian Orogeny, causing, for instance the deformation of early Devonian red beds in the Long Point area of Western Newfoundland (Cawood et al., 1988).

In the British Isles, sedimentation in the Dublin-Northumberland Trough and in the Midland Valley Graben was interrupted during the middle Devonian by sinistral convergent wrench movements, inducing low relief folding of the Lower Old Red Sandstone series, the development of a regional unconformity and cleavage in the Lake District and Wales (Bluck, 1983; Soper at al., 1987, McKerrow, 1988; Woodcock et al., 1988; Ziegler, 1982, 1988). In the Orcadian Basin, on the other hand, unconformity-bound middle Old Red Series consist of some 5000 m of lacustrine shales and fluviatile sands. Their accumulation was accompanied by the extrusion of volcanics and syn-depositional deformations, reflecting the interplay of transpressional and transtensional stresses; these can be related to sinistral movements along the Great Glen fault system (House et al., 1977; Watson, 1985). Post-depositional deformation of these strata involved folding, steep reverse faulting and locally overthrusting, as for instance in the Caithness area of northern Scotland (House et al., 1977) and in the Orkney Isles (Coward and Enfield, 1987). On the other hand, the Hornelen-Solund complex of basins of southern Norway subsided rapidly during the middle Devonian in response to tensional reactivation of Caledonian low angle thrust faults and concomitant sinistral wrench faulting (Steel, 1976; Roberts, 1983; Séranne and Séguret, 1987). These middle Devonian deformations are probably the expression of intensified movements along the Arctic-North Atlantic megashear during the Acadian-Ligerian Orogeny.

In the Trondheims Fjord area (Mid-Norway) sedimentation ceased also during the middle Devonian and the margin of this narrow basin became overthrusted by the Caldonian basement (Roberts, 1983; Steel et al., 1985; Norton, 1986). On the other hand, the Central East Greenland Basin continued to subside, whereby there is evidence for intermittent tensional and compressional, syn-depositional deformations (Hudson Land disturbances). The Mid-Devonian evolution of this basin was accompanied by extrusive and intrusive igneous activity, suggesting that deep crustal fracturing facilitated the ascent of mafic and acidic magmas (Haller, 1971; Henriksen and Higgins, 1976). Cooling ages of metamorphics, furthermore, show that uplift and unroofing of the Caledonian fold belt continued (Rex and Higgins, 1985). In northeastern Greenland the intrusion of the mafic-felsic Midtkap igneous complex, dated as 380 ± 5 Ma, is thought to be related to sinistral wrench movements (Håkansson and Pedersen, 1982; Pedersen and Holm, 1983). In Central Svalbard the accumulation of

continental clastics continued in a fault-controlled basin. Sedimentation terminated at the transition from the middle to the late Devonian. Middle Devonian series shows a marginal marine influence; this leads to the assumption that marine transgressions had reached across the Barents Shelf into the area of the Svalbard archipelago. According to the model of Harland et al. (1984), the latter was not yet assembled to its present configuration by middle Devonian time. The distribution of middle Devonian sediments on the Barents Shelf, as shown in Plate 5, is conjectural. Reflection seismic data suggest, however, that individual grabens and troughs had already started to subside during this time (Rønnevik and Beskow, 1983; Rønnevik and Jacobsen, 1984).

3.3. Inuitian Fold Belt and Franklinian Basin

Middle Devonian progressive uplift of the Inuitian fold belt, in response to continued convergence of the Arctic and Laurussian cratons, is reflected by increased clastic supply to the Franklinian Basin.

During the Eifelian, the deeper parts of this basin were still rimmed to the south by the carbonate platform of the Bird Fjord formation, that prograded basinward. This process was impeded and finally terminated by an increasing influx of clastics form the rising Inuitian Orogen to the north. Shallowing-out of basinal areas, the result of accumulation of deeper water clastic fan deposits, was followed during the late Eifelian and Givetian by the southwestward progradation of deltaic complexes onto the stable shelf. At the same time the carbonate bank edges receded stepwise southward in response to cyclical relative sea-level rises and an increase in clastic influx. During late Givetian the north-eastern parts of the Canadian Arctic Archipelago were occupied by a broad alluvial plain, giving laterally way to the west and southwest to marine-deltaic and pro-delta complexes. By this time the carbonate bank edge had receded to the Great Slave Lake area (Slave Point-Presqu'ile barrier reef complex) whereby the reefal Key Scarp platform (Great Bear Lake area) formed an isolated outlier (Bassett and Stout, 1967; Embry and Klovan, 1976; Smith and Stearn, 1982; Balkwill and Fox, 1982). During the middle Devonian the Inglefield uplift ceased to exist as a separate tectonic element and became buried beneath the Inuitian molasse sequence (Smith and Okulitch, 1987). Similarly, tectonic activity decreased during the middle Devonian along the Boothia Arch (Kerr, 1977) that remained, however, a positive feature influencing sedimentation patterns.

The middle Devonian evolution of the Franklinian Basin conforms closely to that of a classical foreland basin. The fact that on the Alaska North Slope Block a carbonate platform was maintained through middle Devonian times (Skajit limestone; Tailleur et al. 1967) illustrates that the western parts of the Inuitian fold belt had not yet been elevated to the same degree as its eastern parts.

On the northern margin of the Inuitian fold belt evidence for the development of a conjugate foreland basin is limited. On Wrangel Island middle Devonian sediments are missing, possibly due to erosion, whilst mixed carbonate-clastic series occur to Chukotka. These clastics were presumably derived from the rising Inuitian Orogen. On Henrietta Island tuffaceous clastics are overlain by a basalt yielding an age of 375 Ma. On the New Siberian Islands middle Devonian series are developed in a mixed carbonate-clastic facies (Vinogradov, 1969; Fujita and Cook, 1986; Gramberg et al. 1986). Although control points are sparse, the above suggests that the northern, Arctica foredeep of the Inuitian fold belt was much narrower and received considerably less clastics than its southern, North American counterpart.

In exposed parts of the Inuitian fold belt, on northern Ellesmere Island, the Mid-Devonian mafic and ultramafic Cape Fanshawe Martin intrusion (383 ± 10 Ma) is roughly contemporaneous with the Midtkap intrusion of northern Greenland (Trettin et al., 1987; Pedersen and Holm, 1983). It is likely that this igneous activity was induced by sinistral wrench movements in the prologation of the Arctic-North Atlantic mega-shear. In this respect, the postulated middle Devonian large scale translation between Laurentia-Greenland and Fennosamatia coincides with the rapid uplift of the western parts of the Inuitian fold belt, and the infilling of the Franklinian Basin with molasse-type clastics. This suggests that this important phase of the Inuitian Orogeny is the combined effect of the slow convergence of the Arctic and the Laurentia-Greenland cratons and of convergent wrench movements at the northern termination of the Arctic-North Atlantic mega-shear.

3.4. Hercynian Geosynclinal System

In the Appalachian foreland basin the Acadian Orogeny is expressed by a phase of rapid subsidence of its proximal (orogen-ward) parts and the out-building of prodelta series and the deltaic Catskill red beds from the rising orogen. At the same time the edges of the carbonate shelves receded craton-ward (Oliver et al., 1967; Donaldson and Shumaker, 1981; House, 1983). The extent to which the Ouachita-Marathon geosynclinal system, forming the southern continuation of the Appalachian geosyncline, was affected by the Acadian Orogeny is uncertain. Moreover, as the present day northeastern margin of the Appalachian basin is erosional, it is uncertain whether this foreland basin extended during the middle

Devonian into the Gulf of St. Lawrence. In the Anticosti Basin corresponding sediments are absent, possibly due to erosion.

In the New England States and the Canadian maritime provinces, where the Acadian fold belt has not been substantially overprinted by the late Carboniferous-early Permian Alleghanian diastrophism, it is evident that the Acadian Orogeny gave rise to intense deformations, metamorphism and an extensive middle to late Devonian plutonism (Williams and Hatcher, 1983; Chorlton and Dallmeyer, 1986; Dallmeyer and Keppie, 1987). Consolidation of the Acadian fold belt was accompanied by regional sinistral shear and local dextral shear movements (Keppie et al., 1985; Keppie and Dallmeyer, 1987). Yet, as much of the Avalon-Meguma terrane is now burried beneath the Mesozoic and Cenozoic series of the Nova Scotia and Newfoundland shelves, it is difficult to assess the extent to which this cratonic block became deformed during the Acadian Orogeny. Neo-autochthonous continental red beds began to accumulate in wrench-induced intra-montane basins (e.g. Magdalen Basin) during the late Givetian (Howie and Barss, 1975; Fralick and Schenk, 1981). This permits dating of the Acadian Orogeny as broadly late Emsian to intra-Givetian.

In the Variscan geosynclinal system, the Ligerian Orogeny was accompanied by the stacking of major, south vergent, basement involving nappes in the Massif Central of southern France (Cogné and Lefort, 1985; Matte, 1983, 1986), and in Galicia (northwest Spain) by the emplacement of east-verging, ophiolite nappes (Martinez-Garcia, 1972, 1980; Iglesias et al. 1983). In Central Iberia, Givetian emplacement of east verging nappes was followed by important dextral wrench movements (Chacón et al., 1983). This was coupled with an increase in clastic supply to the Asturo-Leonese Basin (Carls, 1983, 1988; Carls and

Lages, 1983). The Aquitaine-Cantabian block became, however, little deformed during the middle Devonian and continued to be occupied by extensive carbonate-clastic and carbonate shelves (Julivert et al., 1983; Kullmann et al., 1982). There is only limited evidence for Ligerian deformations in the area of the Western and Central Alps corresponding to the possible Intra-Alpine Terrane (see Ziegler, 1986, 1988).

Middle Devonian reactivation of the northern margin of the Ligerian-Moldanubian Cordillera is evident by the shedding of Givetian molasse-type clastics into the Central Armorican Basin and of turbidites onto the Barrandian carbonate platform (Saxothuringian Basin, sensu latu). Moreover, in the Saxothuringian Basin medium-pressure metamorphics testify to early middle Devonian orogenic activity involving north and south vergent thrusting (Franke, 1988).

In the Rhenohercynian Basin, compressional stresses temporarily overcame back-arc extension, at least in its western parts, and induced a first phase of closure of this oceanic back-arc basin. This was accompanied by the shedding of turbidites form the Normannian High into the western parts of the Rhenohercynian Basin (Rolet et al., 1986; Holder and Leveridge, 1986; Floyd and Leveridge, 1987). In the central parts of the Rhenohercynian basin alkaline-bimodal rift volcanism persisted, however, through middle Devonian times. This suggests that back-arc extension, governing the evolution of the Rhenohercynian Basin, did not become fully overpowered by back-arc compressive stresses, as was the case in the Central Armorican-Saxothuringian Basin (Ziegler, 1982, 1988).

It is uncertain whether also the eastern parts of the Variscan geosynclinal system were affected by a Ligerian compressional phase. Adamia et al., (1981) and Samygin and Khain, (1985) visualise for the Caucasus area a complex arc-trench system involving a tensional inter-arc basin. There is apparently no evidence for a major middle Devonian collisional event. The Mid-Devonian subsidence of the Donets Graben on the Sarmatian Platform (Vinogradov, 1969) may be related to back-arc extension.

3.5. Platform Areas

Middle Devonian times correspond to a period of cyclically rising sea levels as evident in North America, Europe, North Africa and on the Fennosarmatian platform (Vail et al., 1977; House 1983; Johnson et al., 1985). This apparently global rise in sea-level, which culminated during the late Frasnian, is likely to be of a tectono-eustatic nature. The contribution of glacio-eustasy to short-term sea-level fluctuations is, however, uncertain due to limited evidence for important middle Devonian continental glaciations in Gondwana (Hambrey and Harland, 1981).

The available palaeomagnetic data suggest that Laurussia remained more or less stationary during the middle Devonian (Fig. 1; Miller and Kent, 1986). The Arctic-North Atlantic Caledonides probably still formed an important topographic feature, acting as a continental divide. On the other hand, the Acadian-Ligerian mountain range, paralleling the southern margin of Laurussia, acted presumably as an important precipitation barrier for seasonal onshore winds. Overall a hot and seasonally arid climate prevailed in much of Laurussia, as evident by the occurrence of red beds and evaporites both on the Moscow Platform and on the Canadian Shelf.

During the middle Devonian, the area of the Old Red landmass became reduced by transgressions, that inundated the Moscow Platform and the shelf areas adjacent to the Appalachian Basin. Furthermore, the surface of the Canadian Shelf Basin became enlarged, whilst much of the Trans-Continental Arch remained above sea-level. Thus the area of the central Laurussian Old Red landmass had by Givetian time dimensions of about 3000 by 3000 km, to which the mountainous areas of the Acadian-Ligerian and the Inuitian fold belts should be added. Clastic supply from the cratonic core of the Old Red landmass to the adjacent shelf basins was substantial in the Franklinian Basin, the Timan-Pechora area, and possibly in the Barents Sea, but was elsewhere rather limited.

On the Fennosarmatian Platform the eustatic rise in erosional base-level induced initially the accumulation of continental clastics; this was followed by the Givetian transgression of shallow, partly evaporitic seas (Vinogradov, 1969; House, 1975). This shelf basin was linked to the west via Poland with the carbonate platforms occupying the northern margin of the Rhenohercynian Basin; these had become established during the Eifelian (see Ziegler, 1982, 1988).

During the Givetian, the Donets Graben on the southern parts of the Fennosarmatian Platform subsided differentially. This was coupled with an elevation of its rift shoulders, corresponding to the Ukrainian and Voronesh highs. On the northern parts of the Moscow Platform the Vyatka graben began to subside at about the same time and in the Timan-Pechora area the Pechora-Kolva rift system came into evidence during the late Eifelian (Ulmishek, 1982). Moreover, the volcanic Kontozero Graben on Kola Peninsula developed during the middle Devonian (Churkin et al., 1981). Further Devonian grabens may occur in the Barents Sea. This highlights the late middle Devonian extension-induced relative tectonic instability of the eastern and southeastern parts of the Fennosarmatian Platform.

On the North American Craton there is no evidence for similar tensional tectonic instability. Transgressions originating from the Appalachian Basin probably linked up to the north via the Moose River Basin and the Hudson Bay (Sandford and Norris, 1973; Sanford, 1987; Dimian et al., 1983; Telford, 1988; Thorp, 1989) with transgressions advancing southward from the Arctic Shelf. Moreover, temporary connections between the Williston and Iowa basins may even have been established during the late Givetian across the Trans-Continental Arch (Braun and Mathison, 1986).

During the late Eifelian and early Givetian the Canadian Shelf was occupied by an extensive carbonate platform. By late Givetian times large parts of this shelf became drowned by a rapid rise in sea-level and the Presqu'ile-Slave Point barrier reef was established at the latitude of the Great Slave Lake. Back reef areas became filled in by the Prairie Evaporites and the Dawson Bay carbonate-evaporite sequence. There is stratigraphic evidence for a late Givetian reactivation of the Peace River-Alberta Arch (Bassett and Stout, 1967). The outlines of the Canadian Shelf Basin, as given in Plate 5, is based on the assumption that erosional remnants of middle Devonian series in the Hudson Bay area (Thorp, 1989) formed part of this basin.

During the middle Devonian extensive carbonate platforms flanked the Trans-Continental Arch and progressively encroached on it in response to rising sea-levels. Reactivation of the Ozark Uplift, prior to the transgression of the late Devonian Chattanouga Shales, is indicated by a major unconformity (Amsden et al., 1967). Similarly the Cincinnati Arch became reactivated at the transition to the late Devonian (Wilson and Stearns, 1963).

4. Late Devonian

The late Devonian framework of Laurussia is summarised by the Famennian palaeotectonic-palaeogeographic map, given in Plate 6. The relative position of Laurussia and Gondwana is illustrated by Plate 1d.

4.1. Plate Boundaries and Continent Assembly

During the late Devonian, Laurussia as a whole remained more or less stationary with the equator crossing by Famennian time the northern parts of the Hudson Bay and Scotland. According to our model, continued sinistral translations between Laurentia-Greenland and Fennosarmatia were paralleled by a gradual narrowing of the Proto Atlantic-Proto Tethys Ocean (Plate 1d). Particularly in the Appalachian system there is no evidence for a major post-Acadian rifting event that could be related to a post-collisional separation of Laurussia and Gondwana, as inferred from palaeomagnetic data (Morel and Irving, 1978; Van der Voo, 1987).

Our model assumes that movements along the Arctic-North Atlantic mega-shear slowed down during the late Devonian and that by the mid-Famennian, the remanent dextral offset between Greenland and Norway had been reduced to some 200 km. Under this scenario broad-scale collision would have occurred between the northern margin of the Barents Shelf and the south-eastern margin of the Arctic Craton during the late Devonian, causing major crustal shortening in the hypothetical Lomonosov fold belt. Furthermore, continued convergence of the Arctic Craton with the northern margin of Laurussia is reflected by accelerated orogenic activity in the Inuitian fold belt which propagated during the late Devonian into the area of the Alaska North Slope (Ziegler, 1969; Bird et al., 1978; Metz et al., 1982; Dutro, 1981).

Along the southern margin of Laurussia, northward subduction of the Proto Atlantic-Proto Tethys Ocean went hand in hand with the clockwise rotational convergence of Gondwana with Laurussia. Initial contacts between these two mega-continents may have been established already during the Famennian in the area of southern Iberia and Morocco. This could explain the wrench-induced subsidence of troughs in northwest Morocco, (Piqué, 1981; Piqué and Kharbouche, 1983), and the concomitant subsidence of the Ougarta Trough extending in a southeasterly direction from the Anti-Atlas into the Sahara Platform (Wendt, 1985; Ziegler, 1988).

The late Devonian evolution of the Variscan geosynclinal system reflects a continued interplay between back-arc extension and compression whereby an important compressional phase straddled the Devonian-Carboniferous boundary (Bretonian Orogeny, Fig. 2; Ziegler, 1982, 1984).

The late Devonian evolution of the Appalachian geosynclinal system, as mirrored by the development of its foreland basin, continued to be governed by compression. In the Canadian maritime provinces Keppie (1985) recognises a late Acadian orogenic pulse straddling the Devonian-Carboniferous boundary.

In the Cordilleran miogeocline tectonic activity increased during the late Devonian and culminated, as reflected by the shedding of synorogenic clastics from western sources, during the late Mississippian in the Antler Orogeny (Gilluly, 1963; Poole et al., 1967; Gabrielse, 1967; Nilsen and Stewart, 1980; Peterson, 1986; Morrow and Geldsetzer, 1988).

During the late Devonian the Siberian and Kazakhstan cratons began to converge with each other. Continued activity along the Sakmarian-Magnitogorsk arc-trench system, causing narrowing of the Ural Ocean, resulted in the collision of the Mugodjarian terrane with the former in Famennian time. This was coupled with the termination of sea-floor spreading in the Sakmarian back-arc basin (Zonenshain et al., 1984, 1987b; L.P. Zonenshain, personal communication). Whereas rifting activity on the eastern parts of the Fennosarmatian Platform gradually abated, the Donets graben remained active during the late Devonian.

4.2. Hercynian Geosynclinal System

Continued late Devonian orogenic activity in the Appalachian system was accompanied by the intrusion of granitoids in Nova Scotia, ranging in age between 370 and 360 Ma (Reynolds et al. 1981, 1987; Keppie, 1982; Dallmeyer and Keppie, 1987). Also in the United States Appalachians late Devonian orogenic activity is documented by the intrusion of granitoids and metamorphism (Dallmeyer, 1978, 1982; Dallmeyer et al., 1986; Osberg, 1982; Secor et al., 1986). This is cor-

roborated by the late Devonian rapid subsidence of the Appalachian foreland basin.

The late Devonian prodelta and deltaic Chemung-Catskill series, which grade upward into continental red beds, attain in the northern parts of the Appalachian foredeep basin thicknesses of up to 2700 m. In time this clastic wedge spread into the southern Appalachian domain, possibly reflecting a southward progression in the intensity of orogenic activity (Oliver et al., 1967; Woodrow et al., 1973; Donaldson and Shumaker, 1981; House, 1983; Ettensohn, 1985; Scheckler, 1986). The extent to which similar series were deposited in the proximal parts parts of the southern most Appalachian and in the Ouachita-Marathon foredeeps is unknown. Limited radiometric data from west Texas and northern Mexico suggest, however, that also this segment of the Hercynian orogenic system was active during the late Devonian (Cook and Bally, 1975; Handschy et al., 1987).

Toward the foreland, the molasse-type Catskill series give way to the basinal, cyclically organic Chattanooga-New Albany shales and their equivalents. During the Frasnian these spilled over large parts of the stable carbonate platforms, which had occupied the Michigan, Illinois and Central Iowa basins during the middle Devonian. (Collinson et al., 1967; Cook and Bally, 1975). This change in depositional patterns was probably the combined effect of eustatically rising sea levels, regional subsidence in response to tectonic loading of the foreland crust by the advancing orogen (Johnson et al., 1985; Quinlan and Beaumont, 1984; Ettensohn, 1987), and increased influx of fine clastics, that smothered the carbonate banks.

In contrast, the late Devonian evolution of the Variscan geosynclinal system was governed by renewed back-arc extension as evident from alkaline-bimodal volcanic activity in the Rhenohercynian, Central Armorican and Saxothuringian basins. This volcanism abated, however, with the onset of the Bretonian orogenic pulse during the Famennian.

During the Frasnian the northern margin of the Rhenohercyian Basin was occupied by reef-fringed carbonate and carbonate-clastic shelves; particularly their northwestern parts became covered, during the Famennian, by regressive sands derived from the Old Red landmass. Influx of clastics and olistostroms into the southern, deeper water parts of the Rhenohercynian Basin from the Normannian and Mid-German highs increased during the Famennian and is thought to be related to the resumption (accelerations?) of subduction processes along their northern margin during the Bretonian orogenic pulse. Also in the Central Armorican, Saxothuringian and Upper Silesian basins, a similar late Famennian increase in clastic supply from southern sources is evident and testifies to gradually intensified compressional tectonic activity (Behr et al., 1982, 1984; Franke and Engel, 1986; Weber, 1984; Holder and Leveridge, 1986; Rolet, 1983; Rolet et al., 1986; Ziegler, 1982, 1988; Paris et al., 1986).).

During the late Devonian, the Aquitaine-Cantabrian platform was occupied by a metastable carbonate and carbonate-clastics platform. This platform was rimmed on three sides by the arcuate Central Iberian-Ligerian fold belt. In the Asturo-Leonese Basin late Devonian series are developed in a deeper water facies (Carls and Lages, 1983; Oliveira et al., 1986).

It is likely that during the late Devonian the southern margin of Iberia was occupied by an arc-trench system, along which oceanic crust of the progressively narrowing Western Proto Tethys became subducted. In the South Portuguese Basin the accumulation of late Devonian shallow marine and flysch-type clastics was accompanied by an alkaline bimodal volcanism (Oliveira, 1982; Julivert et al., 1983). This can be taken as an indication for back-arc extension behind the postulated South Iberian arc-trench system.

In the Iberian domain, evidence for a Bretonian orogenic pulse is restricted to Galicia (Martinez-Garcia, 1972; Julivert, 1979, 1983), northern Algeria (Bouillin and Perret, 1982) and northern Morocco (Piqué, 1982). The latter areas may have formed part of the South Iberian arc-trench system, which presumably collided during the Famennian with the northern margin of Africa, giving rise to the subsidence of wrench induced basins in northwestern Morocco in which flysch-type clastics accumulated (Piqué, 1981, 1982; Piqué and Kharbouche, 1983; Michard et al., 1982; Aifa, 1987). This initial contact between Laurussia and Gondwana apparently facilitated the exchange of non-marine faunas and floras and thus the development of cosmopolitan biotas (Young, 1987).

Limited radiometric date suggest that also the intra-Alpine domain of the Variscan geosynclinal system was affected by diastrophic events at the Devonian-Carboniferous boundary. On the other hand, the ill-defined Austo-Alpine and South Alpine-Dinarid terranes lack evidence for Bretonian deformation and had presumably not yet collided with the Variscan arc-trench system during the latest Devonian (see Ziegler, 1988).

4.3. Arctic-North Atlantic Mega-shear and Associated Basins

In the Canadian maritime provinces, thick late Devonian continental red beds, that accumulated in a system of wrench and pull apart basins, form part of the post-Acadian neo-autochthonous sequence. The evolution of these basins was accompanied by repeated volcanic activity and syndepositional deformations. Controlling fault systems are the sinistral Lubec-Belle Isle-Cobequit-Cabot fault and the dextral Minas (Chedabucto) geofracture. Basins associated with this fault system are the Fundy, Magdalen and Sidney basins (Howie and Barss, 1975; Fralick and Schenk, 1981; Bradley, 1982; Carter and Pickerill, 1985; Fyffe and Barr, 1986; Keppie, 1985). In northern Newfoundland the Cabot fault splits up into the fault systems which controlled the subsidence of the St. Anthony Basin (Cutt and Laving, 1977).

Parts of this fault system probably linked up with the fault systems associated with the Devonian-Carboniferous basins of the British Isles. In Ireland and in the northern United Kingdom, sedimentation resumed during the late Devonian in the basins that had been subjected to middle Devonian wrench related inversion. Upper Old Red clastics generally attain only moderate thicknesses, but reach a maximum of 1000 m in the Midland Valley Graben, 1500 m in the Orcadian Basin and locally 3000 m in northern Ireland. Late Devonian volcanic activity is restricted to the Orcadian Basin. Sinistral wrench movements probably persisted during the accumulation of the Upper Old Red Series (House et al., 1977; Bluck, 1984). However, it is assumed that the bulk of the middle and late Devonian sinistral translations between Laurentia-Greenland and Fennosarmatia was taken up along a hypothetical fault system located to the west of the Hebrides, with only relatively minor motions taking place along the Great Glen, Highland Boundary and Southern Uplands faults.

In the North Sea, the area of Old Red Sandstone accumulation became rapidly enlarged during the late Devonian to the extent that communications were established between the Or-

cadian and Midland Valley basins and the northwestern margin of the Rhenohercynian Basin. This reflects the progressive degradation of the Caledonian fold belts which had formerly occupied much of the North Sea area.

In the Central East Greenland Basin, accumulation of late Devonian Old Red clastics continued under a tectonically mobile regime. Deformation patterns are indicative of repeated wrench deformations (Haller, 1971; Henriksen and Higgins, 1976). On the other hand, in Svalbard sedimentation became interrupted during the late Devonian Svalbardian tectonism, during which the fill of the Central Spitsbergen Through became compressionally deformed (Lamar et al., 1986; Roberts, 1988). The accumulation of continental series resumed during the latest Devonian under a transtensional setting (Steel and Worsley, 1984).

The late Devonian Svalbardian movements probably contributed substantially to the assembly of the Svalbard archipelago to its present configuration (Harland et al., 1984). These movements were accompanied by a broad uplift of the northern Barents Shelf. This caused the erosion of the former sedimentary cover of Franz Josef Land. At the same time a foreland basin, paralleling the Lomonosov fold belt, may have developed. Deltaic and continental clastics, transported along the axis of this basin, and partly derived from the North Barents Shelf uplift, were shed onto the Severnaya Zemlya carbonate platform (Churkin et al., 1981; Gramberg et al. 1986).

4.4. Inuitian Fold Belt and Cordilleran Miogeocline

The main feature of the late Devonian evolution of the Inuitian fold belt is its rapid propagation into the area of the Alaska North Slope, where carbonate sedimentation terminated during the late Devonian and gave way to the deposition of the synorogenic Kanayut conglomerates. Intense deformation of the earlier series was accompanied by the intrusion of granitoids (Gabrielse, 1967; Bird et al., 1978; Metz et al., 1982; Moore and Nilsen, 1984; Hubbard et al., 1987). Sedimentation patterns, combined with reflection seismic data from the North Slope, suggest that this fold belt had a double vergence.

Deformation of the Alaskan branch of the Inuitian fold belt was paralleled by increased tectonic activity along the outer platform of the Cordilleran miogeocline, as evident by the shedding of clastics from the Cassiar-Yukon-Fairbanks platform (Nation River and Earn graywackes; Gabrielse, 1967; Mortensen, 1982; Morrow and Geldsetzer, 1988; Howell and Wiley, 1987). The occurrence of early Famennian alkaline bimodal volcanics in the Nevadan and Yukon parts of the Cordilleran miogeocline (Dub, 1987; Gordey, 1988) suggests that the area was affected either by a phase of back-arc extension, or by wrench faulting during the early phases of the Antler Orogeny (Eisbacher, 1983; Crasquin et al., 1986). Late Devonian-earliest Carboniferous intensified orogenic activity along the Cariboo-Antler fold belt, extending from Alaska to Nevada, was accompanied by the intrusion of granitoids (Lethiers et al., 1986).

In the Canadian Arctic Archipelago, contemporaneous further uplift of the Inuitian fold belt is indirectly evident by the advance of alluvial fans, that by latest Devonian time had reached the area of Banks Island (Embry and Klovan, 1976). This was paralleled by progradation of the deltaic Imperial clastics across the Anderson Plains and Mackenzie Shelf to the margins of the Richardson Trough. In the latter, turbidites flowed southward into the Cordilleran back-arc basin. Outbuilding of the Inuitian molasse deposits was coupled with a fur-

ther southward displacement of the carbonate bank edge on the Canadian Shelf during the Frasnian. However, regressive carbonates advanced northward again into the area of the Great Slave Lake during the Framennian (Bassett and Stout, 1967; Ziegler, 1969). During the Frasnian the Cornwallis-Boothia arch became accentuated once more (Kerr, 1977). This illustrates that also during the late Devonian orogenic phase of the Inuitian fold belt important compressional stresses were exerted onto its foreland. On northern Ellesmere Island, late Devonian continental clastics were apparently deposited in an intramontane basin as part of a neo-autochthonous sequence (U. Mayr, personal communication).

The northern, Arctic foredeep of the Inuitian fold belt received also during the late Devonian apparently far less clastics than its southern, North American counterpart. On Wrangel Island, a Frasnian section, consisting of sandstones, shales and carbonates, represents the youngest Devonian deposits; Carboniferous clastics overlay unconformably Devonian sediments. On the New Siberian Islands, late Devonian deposits consist of sandstones and carbonates (Fujita and Cook, 1986). On Chukotka, late Devonian shallow marine shales grade laterally into carbonates and shales (Vinogradov, 1969). The palinspastic reconstruction for this area, as given in Plate 6, suggests that a narrow rim of continental and deltaic series paralleled the northern deformation front of the Inuitian fold belt, and that these clastics grade laterally, away from the actual foreland basin, into a carbonate-clastic dominated shelf facing the Proto-Arctic Ocean (Vinogradov, 1969).

4.5. Platform Areas

During the Frasnian sea-levels continued to rise, reached a high stand at the transition from the Frasnian to the Famennian, receded cyclically and reached a temporary low-stand at the Devonian-Carboniferous boundary (Vail et al., 1977; House, 1983; Johnson et al., 1985). The Famennian regression coincides with the development of extensive inland icesheets in South America (Hambrey and Harland, 1981; Caputo and Crowell, 1985; Caputo, 1985).

During the late Devonian, the dimensions of the Old Red landmass did not change substantially. Its northern margin was rimmed by the rising Inuitian-Lomonosov orogen and its southern margin by fold belts of the Hercynian geosynclinal system. The Arctic-North Atlantic Caledonides continued to act as a continental divide separating terrestrial-lacustrine faunal provinces (Greiner, 1978). Palaeomagnetic data suggest that by late Devonian time the Old Red landmass still straddled the equator. The prevalence of red beds in the northern parts of the Appalachian foredeep, in the basins of the Canadian maritime provinces, in Northwest Europe and in Central East Greenland, as well as the occurrence of evaporitic series on the Canadian Shelf and the Moscow Platform, indicate that a hot climate with seasonally restricted rainfall prevailed (Woodrow et al., 1973). Only the coal bearing molasse series of the Franklinian foreland basin testify to a more humid environment in which major stream systems, originating in the Inuitian fold belt, probably played an important role.

The Moscow Platform was occupied during Frasnian and Famennian times by an extensive carbonate-clastic shelf. To the west, this shelf linked up via the Baltic and southern Poland with the carbonate dominated platforms of the northern parts of the Rhenohercynian Basin. Persistent positive areas on this vast shelf correspond to the Belorussian High, the East Silesian Massif and the Ukrainian and Voronesh highs. Late Devonian subsidence of the Donets Graben was punctuated by repeated

volcanic activity. Similarly there is evidence for rift-related volcanism on the Polish carbonate platform (Vinogradov, 1969; House, 1975; Ziegler, 1982, 1988). This tectonic instability may reflect continued activity of back-arc convective systems associated with the north-plunging Proto-Tethys subduction zone.

The carbonate shelf of the Moscow Platform extended northward onto the eastern Barents Shelf and into the area of Taimyr Peninsula and Severnaya Zemlya. Rifting activity in the Vyatka Graben and in the Timan-Pechora area abated during the Famennian, but persisted into the early Carboniferous. Similarly, the Kontozero rift on Kola Peninsula remained active (Vinogradov, 1969; Ulmishek, 1982; Gortunov et al., 1984; Gramberg et al., 1986). The nature of the geodynamic processes governing the evolution of these grabens is not clear. Their initial rifting phases coincide with the onset of back-arc extension in the Sakmarian basin; their later phases could be related to wrench-movements along the Trollfjord-Komagelv fault, transsecting northernmost Norway in an east-west direction; this fault may be considered as forming part of the Arctic-North Atlantic mega-shear system. During the late Devonian clastic influx from the Fennosarmatian High onto the Moscow Platform apparently deceased. Facies patterns shown for the Barents Sea are conjectural and have been extrapolated from very limited control. The shedding of late Devonian clastics onto the Severnaya Zemlya carbonate shelf can be related to the collision of the Barents Shelf with the Lomonosov fold belt.

The Famennian "Condroz Sandstone" regression on the northern Rhenohercynian Shelf reflects the development of new drainage systems in the area of the North Sea area, an increase in clastic influx from the domain of the Arctic-North Atlantic mega-shear and, perhaps only in a subordinate way, a eustatic lowering of sea-levels (House, 1975; Ziegler, 1982, 1988; Paproth et al., 1986).

On the North American Craton, the areal extent of carbonate platforms became reduced during the late Devonian, partly in response to tectonically and eustatically induced cyclical deepening events and partly by increasing clastic influx from the rising Inuitian and Appalachian orogens (Bassett and Stout, 1967; Cook and Bally, 1975; Johnson et al., 1985). The palaeogeographic reconstruction, given in Plate 6, assumes that the Canadian Shelf continued to be in communication with the Appalachian Basin via the Hudson Bay. In the Moose River Basin and the Hudson Bay, Frasnian and early Famennian shallow marine, partly kerogenous shales are the youngest preserved Palaeozoic strata (Thorpe, 1988; Tellford, 1989) and it can only be speculated that late Famennian and possibly also early Mississippian deposits were removed during subsequent erosional events. On the Canadian Shelf, Frasnian reef complexes are overlain by regressive Famennian carbonates. These grade southward, toward the basin edge, into evaporitic and continental clastic series. During the late Devonian the long standing Peace River High-Alberta Arch became finally overstepped (Bassett and Stout, 1967; Burrowes and Krause, 1987).

Similarly, the margins of the Trans-Continental Arch became significantly overstepped during the late Devonian. Along its southern flank carbonate deposition was restricted to narrow shelves, remnants of which are preserved in the northern parts of the Central Iowa Basin. These were offset by sediment starved shale basins occupying the distal parts of the Appalachian-Ouachita-Marathon foredeep (Cook and Bally, 1975).

At the transition from the Devonian to the Carboniferous, compressional intra-plate stresses, related to the Antler Orogeny of the Cordil-

leran miogeocline and to the late Acadian orogenic pulse in the Appalachians, induced a minor uplift of the Trans-Continental Arch, of the Central Montana High, of numerous low relief structures in the northern Rocky Mountain plains and of the Stanbury-Defiance Arch that extends from the Great Salt Lake to New Mexico. At the same time the Cedar Creek and Nesson anticlines in the southern parts of the Williston Basin became upthrusted along steep reverse faults (Sandberg and Maple, 1967; Poole et al., 1967; Cook and Bally, 1975; Clement, 1986; Ross and Ross, 1986; Craig and Waite-Connor, 1979). Similarly, the Cincinnati Arch became repeatedly reactivated by compressional foreland stresses during the middle and early late Devonian; this is evident by a major pre-Famennian hiatus and the occurrence of Chattanooga equivalent shore line sands on the western flank of the Nashville Dome (Wilson and Stearns, 1963; Oliver et al., 1967; Cook and Bally, 1975).

Unlike the Fennosarmatian Platform, which was affected by tensional tectonics during the late Devonian, the North American Craton became apparently deformed by compressional intra-plate stresses during the late Devonian-earliest Mississippian.

On the Sahara Platform, Frasnian time corresponds to a maximum transgression during which open marine, partly kerogenous shales were deposited on its northwestern distal shelf parts. Famennian-Tournaisian times correspond to a regressive cycle. Differential subsidence of the Ougarta Trough and of wrench induced basins in Morocco, reflecting the collapse of the northwestern promontory of the Sahara Platform, may be related to the oblique collision of the latter with the South Iberian arc-trench system. These deformations were accompanied by the accumulation of thick regressive clastics in the Anti-Atlas Trough and of flysch-type series in the Sidi Betach Basin of northern Morocco; furthermore there is evidence for syndepositional alkaline volcanism (Piqué, 1981, 1982; Piqué and Kharbouche, 1983; Michard et al., 1982; Wendt, 1988). New palaeomagnetic data from Morocco suggest, however, that by latest Devonian time northwest Africa was still located a considerable distance to the south of cratonic Europe (Aifa, 1987). On the other hand, it must be realised that the amount of crustal shortening achieved during the Carboniferous in the Variscan fold belt is difficult to quantify and could be considerably larger than assumed in the reconstructions presented here.

5. Early Carboniferous

The late Visean configuration of Laurussia is summarised by the palaeotectonic-palaeogeographic map, given in Plate 7. The suggested position of Laurussia relative to Gondwana is shown in Plate 1e.

5.1. Plate Boundaries and Continent Assembly

During the early Carboniferous, continued subduction of oceanic lithosphere at the Appalachian arc-trench system was paralleled by increasing clastic influx into the Appalachian foreland basin. This was accompanied by progressive narrowing of the Proto Atlantic Ocean, which by late Visean time had probably a remnant width of 600-800 km. During the early Carboniferous, Gondwana converged in a clockwise rotational mode with Laurussia. At the same time Laurussia drifted northward and began to rotate clockwise. The apparent increasing conformity in the drift patterns of Gondwana and Laurussia is suggestive of their progressive coupling. These plate movements were paralleled by the convergence of the Siberian and Kazakhstan cratons and further narrowing of the Ural Ocean (Zonenshain et al., 1987b).

Clockwise rotation and northward drift of Gondwana was coupled with widening of the eastern Proto Tethys Ocean, progressive closure of its western parts and continued subduction processes along its northern margin. In the Variscan geosynclinal system orogenic activity abated apparently during the Tournaisian-early Visean, but accelerated sharply during the late Visean. This marked the onset of the Himalayan-type Main-Variscan Orogeny during which the Austro-Alpine terrane became accreted to the southern margin of Laurussia. At the same time the long standing Rhenohercynian back-arc basin became closed and in its place the Variscan foreland basin developed (Matte, 1986; Behr et al., 1984; Ziegler, 1982, 1986, 1988).

According to the model presented in this paper, the late Devonian remnant dextral offset of some 200 km between Laurentia-Greenland and Fennosarmatia was recovered by late Famennian to Visean sinistral strike-slip motions along the Arctic-North Atlantic megashear. These displacements were taken up in the Inuitian-Lomonosov fold belt. Stratigraphic data indicate that the Inuitian fold belt became consolidated during the early Mississippian Ellemesian orogenic pulse. Its late Missisippian progressive collapse was accompanied by tensional and wrench movement (Trettin and Balkwill, 1979; Kerr, 1981; Balkwill and Fox, 1982).

Termination of the Inuitian Orogeny coincides with the northward drift and clockwise rotation of Laurussia as a whole. This may have involved abandonment of the sea-floor spreading axis in the Proto Arctic Ocean that was active during the Devonian southward convergence of the Arctic Craton with Laurussia.

In the domain of the Ural Ocean, Kazakhstan

collided with the Sakmarian-Magnitogorsk arc-trench system during the middle Visean, and a new east-plunging subduction zone developed along its southwestern leading edge. At the same time convergence between the Siberian and Kazakhstan cratons accelerated (Zonenshain et al., 1984, 1987a, b; L.P. Zonenshain, personal communication). This marked onset of the actual Uralian Orogeny.

Along the western margin of North American Craton orogenic activity intensified during the latest Devonian-early Carboniferous, giving rise to the Antler Orogeny. During this orogenic cycle closure of the oceanic Klamath-Sierran back-arc basin was accompanied by the emplacement of the Roberts Mountain allochthon on the southwestern shelf margin of the Cordilleran miogeocline (Gilluly, 1963; Burchfiel and Davis 1972, 1975; Nilsen and Stewart, 1980; Frazier and Schwimmer, 1987). The Antler Orogeny can be rated to the sinistral oblique convergence of the Pacific Plate with the clockwise northward rotating North American Craton (Plate 1e).

During the early Carboniferous Laurussia had reached its largest dimensions as a result of the final accretion of the Arctic Craton, encompassing the Seward Peninsula, Chukotka and the New Siberian Islands. Large scale collision between Laurussia and Gondwana occurred during the late Visean; with this the process of their suturing, culminating in the assembly of Pangea, had commenced (Johnson, 1981).

5.2. Hercynian Geosynclinal System

A large number of radiometric dates form the Appalachian fold belt indicate that it was affected by a major tectono-thermal event during the Mississippian (Fyffe et al., 1982; Osberg, 1982; Dallmeyer et al., 1986; Dallmeyer and Keppie, 1987; Cook and Bally, 1975). Early

Carboniferous orogenic activity in the Appalachian fold belt was paralleled by accelerated subsidence and cyclical shallowing out of the Appalachian foredeep basin and by repeated advances of deltaic complexes onto the foreland, particularly during the Kinderhookian and Osagean (Roeder et al., 1978; Donaldson and Shumaker, 1981; Scheckler, 1987). During Meramecian times clastic influx from the orogen abated and carbonate platforms were re-established in much of the Appalachian foreland basins; during the Chesterian the supply of clastics from the Appalachian fold belt increased again (Craig and Waite-Connor, 1979); as such, this may testify to an acceleration of orogenic activity. On the other hand, progressive deflection of the foreland crust under the weight of the advancing nappe systems, combined with glacio-eustatic sea-level changes, may have played an important role in clastic dispersal patterns. The occurrence of late Mississippian deltaic deposits in the South Appalachian-Ouachita foreland basin, and of turbidites in the Arkoma basin, indicates that also this segment of the Appalachian orogen was progressively rising during the early Carboniferous (Mack et al., 1983; Southerland, 1988). This suggests that the early Carboniferous evolution of the Appalachian fold belt was governed, as during the late Devonian, by continuous, northwestward subduction of the Proto Atlantic Ocean.

In the Varsican geosynclinal system back-arc rift-related volcanism resumed once more during the Tournaisan and early Visean, but came to an abrupt end by Mid-Visean time. On the other hand, synorogenic clastics continued to be shed throughout early Carboniferous times into the southern parts of the Central Armorican and Saxothuringian basins. Moreover, the influx of clastics from the Normanian-Mid German High into the Rhenohercynian Basin increased gradually during the Tournaisian and early Visean and became massif during the late

Visean-early Namurian. This suggests an inter-play between back-arc extension and back-arc compression during the earlier parts of the early Carboniferous that was followed by the final overpowering of back-arc tensional forces during the late Visean. Progressive, but possibly intermittent, closure of the oceanic Rhenohercynian back-arc basin along a south-plunging subduction zone, paralelling the northern margin of the Normannian-Mid German highs, probably began during the latest Devonian. This process accelerated and became continuous during the middle Visean. Apart from the sedimentary record of the Rhenohercynian Basin this concept is supported by the occurrence of early Carboniferous I-type granites in the Vosges and Odenwald (Holl and Altherr, 1987; Volker and Altherr, 1987; Henes-Klaiber et al., 1989). By early Namurian time oceanic domains in the Rhenohercynian basin had probably been closed and its northern carbonate shelves subsided rapidly under the load of advancing nappes; with this, the basin assumed the geometry of a typical foredeep, in the proximal parts of which synorogenic, flysch-type Culm-series accumulated (see Ziegler, 1982, 1988).

The carbonate-clastic shelf on the Aquitain-Cantabrian platform was maintained during the early Carboniferous, whereas flysch was shed into the narrow foredeeps paralelling the Central Iberian Range (Julivert, 1987) and possibly also the Ligerian Cordillera. In the South Portuguese Basin, accumulation of synorogenic flysch during the early Carboniferous was accompanied by a mixed alkaline and calcalkaline volcanism, possibly related to wrench movements accompanying the northeastward transpression and escape of the Aquitaine-Cantabrian block (Oliveira, 1982). In northern Morocco wrench and rifting activity continued during the early Carboniferous. In time clastic influx from local highs and the rising South Iberian arc system, that was characterised by a

calcalkaline magmatism, increased in importance (Piqué, 1981, 1982; Piqué and Kharbouch, 1983; Wendt, 1988; Beauchamp and Izart, 1987).

Closure of the western Proto-Tethys during the Variscan Orogeny was probably not synchronous and involved multiple collisions between continental fragments. Correspondingly, onset of the Main-Variscan orogenic deformations phase varies in time and space in the different parts of this fold belt. In the Eastern Mediterranean and Black Sea area the Proto Tethys remained, however, open through Carboniferous times; the southeastern margin of Laurussia remained therefore in a Pacific-type subduction setting (Ziegler, 1984, 1988; Zonenshain et al., 1987b).

5.3. Arctic North Atlantic Mega-shear and Associated Basins

In the Canadian maritime provinces, the wrench-related Fundy, Magdalen, Sidney and St. Anthony basins continued to subside during the early Carboniferous; the evolution of these basins was punctuated by repeated intrusive and extrusive igneous activity and was accompanied by syndepositional deformations (Howie and Barss, 1975; Fralick and Schenk, 1981; Bradley, 1982). Late Visean marine incursions into these basins, that gave rise to the accumulation of evaporitic series, originated from the carbonate shelves of Ireland and from the South Portuguese Basin (Ziegler, 1988).

In the British Isles, the early Carboniferous evolution of the Dublin-Northumberland Trough, the Midland Valley Graben and related basins was governed by extensional tectonics, as evident by a widespread alkaline, bimodal rift volcanism (Francis, 1978, 1988; Leeder, 1982; Kirton 1984). Fault systems controlling the subsidence of these basins were essentially the same as those active during the Devonian; therefore it may be assumed that they became reactivated during the early Carboniferous by diverging wrench movements. Significantly, rift-type magmatism persisted in these basins into early Westphalian time, that is, well beyond the extinction of back-arc rift volcanism in the Rhenohercynian Basin. The late Carboniferous tensional evolution of these grabens was paralleled by rifting in the Norwegian-Greenland Sea area (Ziegler, 1982, 1988).

The sedimentary record of the wrench-dominated Central East Greenland Basin terminated with the accumulation of early Tournaisian red beds and lacustrine shales. These sediments are folded and unconformably overlain by late Carboniferous and Permian series that were deposited under a tensional regime. The last phase of deformation of the Devonian-lowliest Carboniferous Old Red Series, referred to as the Ymerland phase, involved thrust faulting and cannot be closer dated than intra-Dinantian (Haller, 1971; Friend et al., 1976; Henriksen and Higgins, 1976).

In western Svalbard, coal bearing continental series, ranging in age from latest Devonian to late Visean, were deposited in a fault bounded, intra-montane, transtensional basin (Steel and Worsley, 1984). At the transition from the Visean to the Namurian, wrench movements, referred to as the Adriabukta phase, locally gave rise to important deformations and a low grade metamorphism (Birkenmajer, 1981). Time equivalent wrench deformations in the northern Greenland fold belt are also associated with low grade metamorphism (Håkansson and Pederson, 1982). At the same time, sedimentation in the Wandel Sea Basin of northeast Greenland, in which early Carboniferous coal bearing continental clastics were deposited under a tectonically unstable

regime, became interrupted (Håkanson and Stemmerik, 1984).

For the Western Barents Shelf, reflection seismic data indicate the late Devonian-early Carboniferous differential subsidence of grabens and troughs, associated with sinistral wrench faults in its western and conjugate dextral ones in its central parts (Rønnevik and Jacobsen, 1984). Contemporaneous activity along the Trollfjord-Komagelv wrench fault, transsecting the coastal areas of northernmost Norway, is indicated by early Carboniferous dyke intrusions (Beckinsale et al., 1975; Johnson et al., 1978).

In summary, there is considerable geological evidence for continued early Carboniferous tectonic activity along the Arctic-North Atlantic mega-shear. With time displacements along it probably decreased in rate and magnitude. The post-Visean evolution of the Norwegian-Greenland Sea area was governed by rifting (Ziegler, 1988).

5.4. Inuitian Fold Belt

The Inuitian fold belt became consolidated during the Ellesmerian orogenic pulse, that terminated in the area of the Alaska North Slope during the early Mississippian. This is suggested by onlap relationship of the Kekituk conglomerates and the Lisburne carbonates against the erosional surface marking the top of Devonian and older folded and thrusted series (Tailleur et al., 1967; Nilsen, 1981; Bird and Molenaar, 1986). From the Brooks Ranges and British Mountains granitic intrusives, ranging in age from 430 to 300 Ma, have been reported. Intrusive activity culminated between 370 and 330 Ma (Gabrielse, 1967; Ziegler, 1969; Dutro, 1981; Bird et al., 1978; Metz et al., 1982; Hubbard et al., 1987). The occurrence of Silurian to middle Devonian granitoids is enigmatic in terms of the stratigraphically dated late

Devonian-early Mississippian Ellesmerian orogeny; these pre-Ellesmerian intrusives have been related by Metz et al. (1982) to intra-continental hotspot activity.

In the Canadian Arctic Archipelago, the accumulation of post-orogenic continental clastics started during the late Visean. Granitoid intrusives with an age range of 345 to 300 Ma have been reported from northern Ellesmere and Axel Heiberg Island (Frisch, 1974; Trettin et al., 1972; Kerr, 1981).

During the Ellesmerian orogenic pulse, the Franklinian basin became scooped out by thin skinned thrust sheets that now constitute the Parry Island fold belt, forming the externides of the Inuitian orogen. During the development of this fold belt, the north-south trending Cornwallis fold belt became overprinted by east-west trending folds, whereas the Boothia Arch became uplifted (Kerr, 1977,1981; Trettin and Balkwill, 1979; Fox, 1985). The thrust front of the south verging Parry Island fold belt describes an arcuate pattern and strikes out to sea in Prince Patrick Island. Its continuation may be found in northwestern Alaska where the Inuitian deformation front loops to the south around the Chukchi Sea Block and Point Barrow. In this area seismic data indicate the presence of north-verging thrusts (Hubbard et al., 1987; see also Grantz et al., 1981). In northeastern Alaska, the Inuitian fold belt underlies the North Slope and is exposed in the British Mountains that straddle the Canadian-Alaskan Border. The Inuitian thrust front is recognised again in the northern Richardson Mountains (SW of Mackenzie Delta) from where its trends in a southwestern direction (Bell, 1973). It is uncertain whether the Alaska North Slope segment of the Inuitian orogen links up to the southwest with the active Cassian-Yuokon-Fairbanks platform or terminates in eastern Alaska as shown in Plate 7 (S. Cameron, personal communication).

To the east the Inuitian deformation front trends across northwestern Greenland into the area of the hypothetical Lomonosov fold belt. On Svernaya Zemlya folded Devonian strata are unconformably overlain by Mesozoic series; this deformation phase may well be early Carboniferous in age and thus equivalent to the Ellemerian orogenic pulse. Gramberg et al. (1986) refer to early Carboniferous granitoids in the eastern parts of Severnaya Zemlya.

The northern deformation front of the Inuitian fold belt is poorly constrained. On Wrangel Island, Carboniferous carbonates and clastics overlay unconformably, Devonian and older folded strata (Fujita and Cook, 1986), thus indicating that it was also affected by the Inuitian Orogeny. Its position within or at the margin of the Inuitian fold belt is, however, uncertain.

Following the early Mississippian consolidation of the Inuitian fold belt, the area of the Canadian Archipelago became uplifted. This was accompanied by the accumulation of continental clastics in northeast to east trending grabens and the extrusion of basaltic flows (Trettin and Balkwill, 1979). Similarly, half grabens, partly wrench induced, subsided in the area of the Alaska North Slope, in which the the Endicott group clastics were deposited. This was coupled with a gradual marine transgression and the re-establishment of carbonate platforms (Tailleur et al., 1967; Dutro, 1981; Nilsen, 1981; Bird and Molenaar, 1986; Hubbard et al., 1987).

On Chukotka and the New Siberian Islands, forming part to the Arctic Craton, early Carboniferous series are developed in carbonate-clastic facies. In the area of the New Siberian Islands, subsidence of the Bel'Kovsk-Nerpalakh Trough was apparently accompanied by the extrusion of basalts (Vinogradov, 1969; Fujita and Cook, 1986; Gramberg et al., 1986).

The post-Ellesmerian tensional and wrench tectonics of the Arctic domain may be related to final movements along the Arctic-North Atlantic mega-shear, to the post-orogenic gravitational collapse of the Inuitian fold belt, and to incipient back-arc extension induced by the decay of the Inuitian subduction zone. The latter presumably governed the late Palaeozoic subsidence of the Sverdrup Basin.

5.5. Platform Areas

During the Visean, the Old Red landmass, occupying the central parts of Laurussia, straddled 20°N latitude. This land area, that probably became substantially enlarged during the early Carboniferous due to regional uplift of the Hudson Bay area, was limited to the north by the inactive Inuitian fold belt and to the south by the rising mountain chains of the active Appalachian-Varsican Orogen. By this time, the relief of the Arctic-North Atlantic Caledonides had presumably been degraded considerably.

Following an apparently global temporary low stand in sea-level at the Devonian-Carboniferous boundary, extensive transgressions inundated large platform areas during the Tournaisian and Visean. Furthermore, the seas advanced via the Rhenohercynian Basin into the area of the Canadian maritime provinces and via the Barents Shelf into the northern parts of the Norwegian-Greenland Sea area (Ziegler, 1988). Climatic conditions became gradually more humid, as evident by the occurrence of the the first coal bearing sequences, for instance on the western Moscow Platform, in the North Sea area and in Spitsbergen.

The Moscow Platform was occupied during the early and middle Tournaisian by an extensive carbonate and carbonate-clastic shelf. A major hiatus, spanning late Tournaisian to early Visean time corresponds to a regional regression that was apparently associated with an accentuation of pre-existing arches. During the middle and late Visean, the Moscow Platform was again occupied by an extensive carbonate and carbonate-clastic shelf containing isolated evaporitic depressions. The Donets Graben continued to subside differentially, during the early Carboniferous. However, volcanic activity apparently ceased by Tournaisian time. This was paralled by the gradual subsidence of its elevated rift shoulders (Vinogradov, 1969; Igolkina and Krivskaya, 1978).

During the early and middle Mississippian, carbonate platforms occupied the southern and western parts of the North American Craton (Cook and Bally, 1975; Craig and Waite-Connor, 1979, Skipp, 1979). Their original extent is, however, difficult to establish due to intense Permo-Carboniferous and younger erosion of these strata. Thus, the interpretation given in Plate 7 is conjectural, particularly as regards the outlines of the Canadian Shelf. However, there is no evidence for a regional Osagian regression that could correlate to the late Tournaisian-early Visean hiatus on the Moscow Platform. Facies developments in northwestern Canada indicate that clastic influx from the Inuitian fold belt decreased rapidly during the Mississippian whereas the supply of clastics from the Cassiar-Yukon Platform into the Cordilleran miogeocline persisted (Gabrielse, 1967; Ziegler, 1969; Mortensen, 1982; Gordey, 1988).

Overall, on the North American Craton, early and middle Mississippian times appear to correspond to a period of only moderate tectonic activity. During the late Mississippian (Chesterian), compressional intra-plate deformations intensified, causing, amongst others, broadscale uplift of the Trans-Continental Arch and a regional regression; the resultant regional unconformity corresponds to the Kaskaskia-Absaroka sequence boundary of Sloss (1984). On the other hand, the Appalachian-Ouachita-Marathon foreland basin and the Cordilleran miogeocline continued to subsided rapidly during the late Mississippian (Eardley, 1951; Sloss, 1963, 1984; Cook and Bally, 1975; Craig and Waite-Connor, 1979; Peterson, 1986; Kluth, 1986; Ross and Ross, 1986).

On the Sahara Platform the latest Devonian-earliest Carboniferous regression was followed by a regional transgression which persisted into the early late Carboniferous (Conrad et al., 1987). During the Visean, much of Northwest Africa was covered by a carbonate-clastic platform, in part containing evaporites.

In North Africa there is no evidence for a Kaskaskia-Absaroka equivalent sequence boundary, as recognised on the North American Craton, nor for a major pre-Visean regression, as evident on the Moscow Platform. However, with the development of increasingly larger continental ice sheets in Gondwana (Hambray and Harland, 1981), it is likely that glacio-eustatic sea-level changes gained in importance during the early Carboniferous. Yet, major intra-early Carboniferous sequence boundaries, evident on the North American craton and on the Moscow Platform, cannot be correlated with each other and are therefore not of a global eustatic origin but must be related to intra-plate deformations.

6. Late Carboniferous

The Namurian and Westphalian configuration of Laurussia is summarised in Plates 8 and 9. The suggested position of Laurussia relative to Gondwana, Siberia and Kazakhstan is shown in Plates 2a and b.

6.1. Plate Boundaries and Continent Assembly

During the late Carboniferous, Gondwana continued to converge in a clockwise rotational mode with Laurussia. Increased coupling between these mega-continents along their mutual collision front is reflected by a clockwise rotation of Laurussia. These plate movements were paralleled by the Namurian collision of Siberia and Kazakhstan, the progressive closure of the Sakmarian back-arc Ocean and the intra-Westphalian collision of Kazakhstan with the passive, southeastern margin of the Moscow Platform.

Clockwise rotation of Gondwana, causing a further widening of the eastern Proto-Tethys Ocean, was accompanied by rapid closure of its western parts and progressive narrowing of the Proto-Atlantic Ocean. During the Namurian and Westphalian, the collision front between northwest Africa and southern Europe propagated eastwards, whereby the South Alpine and Dinarides terrain(s) became incorporated in the Variscan fold belt. During the Namurian and Westphalian main phases of the Variscan Orogeny, major north and south verging, basement cored nappes were emplaced in

Western and Central Europe. Increasing space constraints caused the northwestward escape of the Aquitaine-Cantabrian block. By late Westphalian time crustal shortening terminated in the Variscan fold belt (Ziegler, 1982, 1988).

The timing of closure of the Proto-Atlantic Ocean and of the initial collision of Gondwana with the Marathon-Ouachita-Appalachian arc-trench system, is poorly constrained, but is thought to have occurred during the late Namurian to early Westphalian. This marked the onset of the Himalayan-type Alleghanian orogenic cycle, that culminated in the early Permian consolidation of the Marathon-Ouachita-Appalachian-Mauretanides fold belt (Osberg, 1983; Rast, 1984; Hatcher, 1985).

In the area of the Norwegian-Greenland Sea and on the southwestern Barents Shelf, rifting commenced during the Namurian. This was paralleled by dextral wrench faulting between northeast Greenland and Spitsbergen, the rapid subsidence of the Sverdrup Basin and the development of the Arctic carbonate shelf (Ziegler, 1988). By Westphalian time the northern and eastern shelves of Laurussia were probably occupied by a coherent carbonate platform.

In the Uralian domain, the Siberian craton collided with Kazakhstan during the Namurian whereby the latter continued to converge with the eastern margin of Laurussia, thus inducing a narrowing of the Sakmarian Ocean. During the Bashkirian-Moscovian (early Westphalian),

Kazakhstan collided with the southern parts of the East European passive margin (Nalivkin, 1982; Artyushkov and Baer, 1983; Zonenshain et al., 1984, 1987a, b).

The late Carboniferous megatectonic setting of the western margin of the North American craton is unclear. Following the Mississippian Antler Orogeny, its evolution appears to have been dominated by back-arc extension and the development of the intra-oceanic Sonoma arc-trench system (Burchfiel and Davies, 1975). This was presumably accompanied by sinistral translations between the North American Craton and the Pacific Plate, as suggested by the palaeomagnetically documented clockwise rotation of Laurussia. Whether the Sonoma arc-trench system linked up to the north with an evolving arc-trench system paralleling the northern margin of Laurussia, and via this with the Uralian subduction system, as shown in Plate 8, is uncertain. However, such a new plate boundary may have developed in conjunction with the late Carboniferous rotational movement of Laurussia.

6.2. Variscan Orogen

In the Variscan segment of the evolving Hercynian megasuture, orogenic activity peaked during the Namurian and early Westphalian and terminated at the transition from the Westphalian C to D.

The framework of the Variscan fold belt can only be unraveled in extra-Alpine areas; in the Alpine and Mediterranean domains available palinspastic reconstructions are of a tentative nature. Thus the representation of the Variscan fold belt in these areas, as shown in Plates 8 and 9, should be regarded as schematic.

The amount of crustal shortening achieved during the Namurian and Westphalian across the Variscan fold belt is difficult to estimate.

Major nappe structures, in part involving basement , occur in the Moldanubian area, in the southern Massif Central, in Iberia and in the Variscan externides (Thiele, 1977; Julivert et al., 1977; Brug et al., 1984; Meissner et al., 1981; Behr et al., 1982; Matte, 1983, 1986; Cazes et al., 1985; Raoult, 1986).

The late Carboniferous apparent polar wander paths (APWP) of Africa and Europe are still insufficiently constrained to assess the total amount of crustal shortening that was achieved between them during the Variscan Orogeny (Daly and Irving, 1983; Aifa, 1987). On the other hand, palaeomagnetic data suggest that the post-Devonian crustal shortening between the Armorican Massif and the northern Variscan foreland does not exceed the palaeomagnetic margin of error and thus cannot be larger than 500 km (Jones et al., 1979). Yet, the widespread occurrence of low-pressure metamorphics and late- to post-orogenic calc-alkaline intrusives in the Variscan fold belt can be taken as indirect evidence that significant amounts of crustal shortening, accompanied by crustal delamination, subduction, and anatectic remobilisation of lower crustal material and partial melting of upper mantle material took place during the Variscan Orogeny (Zwart and Dornsiepen, 1978; Matte, 1983, 1986; Francis, 1988; Ziegler, 1984, 1986).

The arcuate configuration of the Variscan fold belt, that involves major oroclinal bending of nappes and wrench faulting, resulted from its draping around a number of Gondwana-derived allochthonous microcratons (for instance, Aquitaine-Cantabrian block, Barrandian, East Silesian massifs), from the configuration of the Devonian-Early Carboniferous back-arc rifts that formed a system of internal successor basins, and from the shape of the tensional Rhenohercynian precursor basin of the northern Variscan foreland basin (Ziegler, 1986, 1988).

During the latest Visean-earliest Namurian Sudetic or Variscan main orogenic pulse, the oceanic domains of the Rhenohercynian back-arc basins became subducted, as indicated by the occurrence of I-type granites in the northern Vosges and Oldenwald (Holl and Altherr, 1987; Volker and Altherr, 1987; Henes-Klaiber, 1989). Closure of this basin, and the ensuing earliest Namurian collision of the advancing nappe systems with the northern shelf of the Rhenohercynian Basin, was accompanied by a phase of block faulting that caused the disruption of its Visean carbonate platforms. Subsequently these platforms drowned in response to tectonic loading. Development of the Variscan foreland basin went hand-in-hand with the accumulation of thick synorogenic molasse and flysch series in its proximal parts; this prograding clastic wedge was offset by a deep water trough in which thick shales were deposited. At the same time clastic influx from the distal basin margins increased (Ziegler, 1982).

The late Visean-early Namurian development of the Variscan foreland basin was paralleled by the folding and partial destruction of the long-standing Central Armorican and Saxo-thurinigian basins (Pfeiffer, 1971; Schmidt and Franke, 1975; Behr et al., 1984). Subsequently, Namurian and Westphalian continental clastics accumulated in local intramontane basins; these sediments form part of the late Palaeozoic neo-autochthonous series of the Variscan Internides (Lützner and Schwab, 1982; Ziegler, 1988).

On the Aquitaine-Cantabrian block, a mixed carbonate-clastic shelf was maintained during Namurian and Westphalian times. This shelf was fringed by the arcuate foreland basin of the Armorican-Central Iberian orocline, in which initially, flysch and, later, paralic coal-measures were deposited. However, the central Cantabrian carbonate platform became smo-thered by clastics and scooped out by thin-skinned thrust sheets during the late Westphalian (Eichmüller and Seibert, 1984; Julivert, 1983; Pérez-Estaún et al., 1988). Incorporation of the South Alpine-Dinarid terrane into the Variscan fold belt, dated as late Westphalian, was accompanied by rapid subsidence of its carbonate platform and the accumulation of synorogenic flysch (Castellarin and Vai, 1981; Vai and Cocozza, 1986).

During the late Namurian and early Westphalian, progressively larger parts of the Variscan foreland basin became dominated by a paralic depositional regime with clastics derived from the rising orogen, as well as from the northern basin margin. This foreland basin extended from the Black Sea through Poland, northern Germany and the British Isles to southwest Portugal, where turbiditic series extend upwards into the Westphalian. Faunal evidence suggests that the western parts of the Variscan foreland basin were in communication with the rift-wrench basins of the Canadian maritime provinces (Trueman, 1946).

In the Variscan foreland basin, Westphalian coal-measures reach thicknesses of up to 3500 m. Cyclical, short-lived marine incursions into this basin, originating presumably from the Moscow platform, are related to glacio-eustatic sea-level fluctuations (Allen and Dineley, 1988).

During the late Westphalian, the proximal parts of the Variscan foreland basin became scooped out by thin-skinned thrust sheets and its southern parts were overridden by basement-involving nappe complexes, for instance in the Ardennes (Casez et al., 1985). At the same time intra-plate compressional stresses caused partial inversions of the Devonian and Carboniferous grabens in the British Isles (Anderton et al., 1979; Ziegler, 1982). Furthermore, these stresses may have impeded the southward propagation of the Norwegian-Greenland Sea rift system that came into evidence during the latest Visean and early Namurian (Steel and Worseley, 1984; Vischer, 1943; Ziegler, 1984). In the Variscan externides the youngest thrust movements are dated as late Westphalian.

6.3. Appalachian Orogen

In the Marathon-Ouachita-Appalachian segment of the Hercynian megasuture, continued northward subduction of the Proto-Atlantic Ocean culminated in its closure and the collision of the West African-Florida-northern South American passive margin with the respective arc-trench systems. However, timing of this collisional event, marking the onset of the Alleghanian orogeny, is poorly constrained and has been estimated variously as having occurred at the transition from the early to late Carboniferous (Spariosu et al., 1984; Secor et al., 1986), or during the Westphalian (Rast, 1984; Keppie, 1985; Ross and Ross, 1985).

In the northern Appalachians of the New England States, the latest Namurian to early Westphalian development of the synorogenic, intra-montane Narragansett Basin speaks in favour of an intra-Namurian onset of the Alleghanian orogenic cycle (Skehan, 1983; Skehan and Rast, 1983; Wintsch and Sutter, 1986). This is in keeping with the geochronological record of the southern Appalachians (Dallmeyer et al., 1986). Whether the Alleghanian collisional event was more or less synchronous along the entire Marathon-Ouachita-Appalachian system is, however, uncertain (Rast, 1984; Osberg, 1982; Hatcher, 1985; Pindell and Dewey, 1982).

During the early Namurian (Chesterian), the proximal parts of the Ouachita-Appalachian fore-deep basin subsided rapidly under the load

of the advancing thrust sheets. Chesterian deltaic and continental series attain thicknesses of over 1000 m in the Appalachian foreland basin and give craton-ward way to mixed carbonate clastic shelves, whereas in the Marathon-Ouachita foredeep some 1000 m to 1500 m thick deeper water clastics are offset to the north by basinal shales and carbonate/clastic shelves. Contemporaneous tectonic instability of the American craton is reflected by the up-warping of the Transcontinental Arch, the accentuation of the Ozark Uplift and the reactivation of the Cincinnati Arch and Nashville Dome. Moreover the Anadarko and Delaware basins subside differentially (Craig and Waite-Connor, 1979; Wuellner et al., 1986).

The end of the Mississippian (early Namurian) corresponds to a period of regional regression that may be related to large scale intra-plate compressional lithospheric deflections rather than to an eustatic low-stand in sealevel. The resulting regional unconformity corresponds to the Kaskaskia-Absaroka sequence boundary (Sloss, 1963, 1984; Frazier and Schwimmer, 1987; see also Cook and Bally, 1975).

During the late Namurian and Westphalian (Morrowian to Desmoinesian), the axis of the Marathon-Ouachita-Appalachian foreland basin migrated craton-ward under the load of the advancing nappe systems. Increased influx of clastics from the rising orogen, and their basin-axial dispersal, resulted in the shallowing out of the Ouachita foreland basin during the Desmoinesian. At the same time deltaic clastics advanced as a broad apron westward from the Appalachian foreland across the Mississippi shelf into Iowa and Missouri, whereas a regional transgression inundated again the area of the Transcontinental Arch. Early and middle Pennsylvanian paralic clastics reach thicknesses of some 2000 m in the proximal parts of the Appalachian foredeep. In the Arkoma Basin (Ouachita foredeep), time equivalent strata are some 8000 m thick (McKee and Crosby, 1975; Southerland, 1988). The Ouachita orogen consisted apparently of a system of tectonically active highlands and intervening successor and inter-arc basins. For instance, deep wells in southern Arkansas and northern Louisiana encountered Desmoinesian to Wolfcampian marine clastics and carbonates that unconformably overlay probable Mississippian and early Pensylvanian volcanics (Frazier and Schwimmer, 1987; Wuellner et al., 1986; R.L. Nicholas, personal communication). The outlines of these basins are not known and are not shown on Plates 8 and 9 that give only a schematic representation of the Marathon-Ouachita Orogen.

Increased coupling between this evolving orogen and the foreland is reflected by the uplift of the ancestral Rocky Mountains; these consist of a system of upthrusted basement blocks, such as the Uncompahgre, Frontrange, Sierra Grande uplifts of Colorado and the Amarillo-Wichita uplift of New Mexico and Oklahoma (Kluth, 1986; Peterson and Smith, 1986; McKee and Crosby, 1975; Ross and Ross, 1985). Uplift of these blocks was coupled with wrench-induced differential subsidence of adjacent basins, such as the Paradox Basin, the Central Colorado Trough, the Anadarko and Palo Duro basins. Apart from these high relief features, a number of lower relief, in part fault bounded, anticlinal axes came into evidence in Montana, the Midcontinent area and in the Michigan and Illinois basins. Most of these intra-plate features started to develop during the Morrowian (late Namurian) and remained active during the Westphalian (Atokan and Desmoinesian) and Stephanian-Autunian (Missourian to early Permian) (McKee and Crosby, 1975; Kluth, 1986).

This is compatible with the concept that broadscale collision between South America and the Marathon-Ouachita arc-trench system occurred during the late Namurian to early Westphalian.

Faunal evidence suggests that during the late Carboniferous the basins of the Canadian maritime provinces were separated from the Appalachian foreland basin by a persistent land barrier (Trueman, 1946; Paproth, 1978). This land barrier may have become tectonically accentuated during the Westphalian by transpressional reactivation of the Acadian deformation front and the uplift of the Adirondack Arch. Intra-Westphalian deformations are also evident in the Magdalen Basin of the Canadian maritime provinces, where they probably involved dextral movements along the Minas geofracture (Keppie et al., 1985).

On the conjugate Gondwana margin of the Appalachian Orogen, the timing and duration of the Hercynian Orogeny is poorly constrained due to the absence of a synorogenic foreland sedimentary prism and the almost total destruction of the externides of the Mauretanides during post-Hercynian times. As the Mauretanides fold belt consists of stacked, basement involving nappes (Lécorché and Sougy, 1978; Lécorché, 1983; Roussel et al., 1984), it is likely that during their emplacement tectonic loading of the foreland caused the development of a typical fore-deep basin. Timing of thrusting can be dated stratigraphically not closer than post-Devonian and geochronological data suggest a late Westphalian to early Permian age (± 300 Ma to 280–270 Ma; Lécorché and Clauer, 1983; Roussel et al., 1984; Dallmeyer and Villeneuve, 1987).

The stratigraphic record of the western Sahara Platform indicates increasing tectonic instability during the Namurian. Early Namurian carbonates and clastics grade upward into evaporitic redbeds and finally into continental clastics. At the same time the Anti-Atlas Arch and the El Biot High became uplifted while inversion of the Ougarta Trough continued. However, on the northeastern Sahara Platform, accumulation of carbonate/clastic shelf series continued into the Westphalian, whereby the rising El Biot Axis formed a barrier to the westward advance of transgressions.

In the Reggane, Tindouf and Taoudeni basins, continental Westphalian redbeds were deposited in tectonically silled basins that became separated by continued upwarping of the Anti-Atlas and Reguibat arches, and the inversion of the Ougarta Trough. Whether these redbeds graded laterally into deltaic series in the postulated Mauretanides foreland basin, as suggested in Plate 9, is uncertain.

Similar to the North American Craton, the northwestern part of the Sahara Platform apparently became increasingly affected by intraplate compressional stresses. During the early Namurian these stresses can be related to increasing coupling between the Variscan Orogen and its southern foreland and, during the late Namurian-early Westphalian, to the collision of the Northwest African passive margin with the Appalachian arc-trench system.

The Namurian and Westphalian palaeotectonic setting of the North Venezuelan margin is uncertain. During the Pennsylvanian, the northern margin of the Guayana Shield was apparently occupied by a stable platform on which fluvial and deltaic clastics, grading northward into carbonates, were deposited. This series, which outcrops in the Sierra Perija and Meride Andes, rests unconformably on Devonian and Siluro-Ordovician strata. The latter are intruded by granites ranging in age from late Cambrian to early Devonian (Juana et al., 1980). Whether this magmatism is of an orogenic or anorogenic nature is not evident from the available literature. The Pennsylvanian clastic and carbonate shelf series of northern Venezuela may be interpreted as forming the distal parts

of an evolving foreland basin, that was subsequently destroyed during the late phases of the Hercynian Orogeny.

6.4. Norwegian-Greenland Sea Rift and Arctic-Barents Shelf

Following the last, intra-Visean transpressional deformations along the Arctic-North Atlantic mega-shear, the late Carboniferous evolution of the Norwegian-Greenland Sea area and of the western Barents Shelf was governed by crustal extension.

In Central East Greenland, Namurian to early Stephanian, up to 1500 m thick continental clastics accumulated in a system of north-south trending half grabens (Perch-Nielsen et al., 1972). Seismic data from the Mid-Norway shelf suggest that also in this area thick late Palaeozoic clastics, contained in a fault bounded basin, underlay Permo-Triassic sediments (Bukovicz and Ziegler, 1985). Seismic data, in part calibrated by wells, show that in the area of the Western Barents Sea the Tromsø-Bjørnøya and Nordkap grabens subsided differentially during the Namurian and Westphalian. Evaporites contained in these rifts are offset by clastics grading upward into carbonates. Contemporaneous rifting is evident on Kola Peninsula (volcanic Kontozero Graben).

Late Carboniferous crustal extension across the Norwegian-Greenland Sea rift was taken up to the north by dextral oblique slip movements between northeastern Greenland and Spitsbergen. In western Svalbard, thick Namurian clastics accumulated in a complex graben system. Wrench-induced partial basin inversion, at the transition from the Namurian to the Westphalian, was followed by a regional transgression and the establishment of an open marine carbonate shelf; however, tectonic activity along discrete fault systems persisted (Steel and Worsley, 1984; Heafford, 1988; Ziegler, 1988).

The Northeast Greenland-Spitsbergen oblique-slip zone extended apparently northwestward into the area of northern Ellesmere Island and the Chukchi Sea, where there is evidence for uplifting of fault blocks and the accumulation of evaporites in down-faulted basins.

Rapid degradation of the Inuitian fold belt, and subsidence of the Sverdrup Basin under a tensional setting, went hand in hand with a regional late Namurian transgression and the establishment of an extensive carbonate shelf occupying much of the Alaska North Slope, Chukotka and the New Siberian Islands (Churkin and Texler, 1981; Hubbard et al., 1987; Fuijta and Cook, 1986; Vinogradov, 1969). In the axial parts of the differentially subsiding Sverdrup Basin, evaporites accumulated during the latest Namurian and early Westphalian (Bashkirian); these are overlain by late Westphalian deepwater shales. These basinal series are offset by reef-rimmed carbonate shelves (Balkwill and Fox, 1982; Stephenson et al., 1987).

Based on the occurrence of continental clastics in Svalbard and on Franz Josef Land, it is assumed that the Lomonosov High remained emergent during the Namurian. It became, however, inundated during the Westphalian transgression, as evident from the establishment of an open marine carbonate shelf in eastern Svalbard and the occurrence of carbonate clasts in the Triassic series on Franz Josef Land. Permian uplift of the Lomonosov High may have caused the truncations of its postulated former late Carboniferous carbonate cover (Heafford, 1988).

By late Westphalian time, probably much of the Arctic and Barents Shelf was occupied by a

more or less coherent carbonate platform. Westphalian uplift of a positive area in the southern parts of Chukotka is suggested by the shedding of clastics onto its northern parts (Vinogradov, 1969). It is speculated that this uplift may be related to the development of a subduction zone along the northern margin of Laurussia.

6.5. Cordilleran Miogeocline

The analysis of the late Carboniferous evolution of the Cordilleran miogeoclinal system is severely hampered by an incomplete and in part only scanty stratigraphic record.

In the Western United States, the Antler Orogeny came apparently to a close during the late Mississippian. The subsequent evolution of the southern Cordilleran area was governed by tensional tectonics, the degradation of the Antler Highland, and the re-establishment of carbonate platforms. This change in the regional stress pattern is thought to be related to the development of the intra-oceanic Sonoma arc-trench system that paralleled the western margin of the North American Craton (Burchfiel and Davis, 1975; Nilsen and Stewart, 1980; Speed and Sleep, 1982; Frazier and Schwimmer, 1987; Miller, 1987).

For the Canadian part of the Cordilleran miogeocline, alkaline intrusions in the Yukon Territory and in southern British Columbia suggest that also this area was dominated by tensional tectonics during the late Mississippian (Mortensen, 1982; Ziegler, 1969). In the external parts of the southern Canadian Cordillera, Pennsylvanian series are developed in shallow marine carbonate-clastic facies. Deltaic and turbiditic sands of Pennsylvanian age occur in the front ranges of the southernmost Mackenzie Mountains. The Cassiar-Yukon-Fairbanks platform, forming the outer margin of the Cordilleran miogeocline, presumably stayed elevated during the Pennsylvanian (Gabrielse, 1967). The carbonate platform of the Alaska North Slope gives way southward to basinal shales (Hubbard et al., 1987).

In view of the above, the late Carboniferous palaeogeographic-palaeotectonic configuration of the Cordilleran margin of the North American Craton, as shown in Plates 8 and 9, should be considered as conceptual.

6.6. Platform Areas

During the Namurian and Westphalian, Laurussia rotated in a counter-clockwise mode by some 14° and 12°, respectively, around a pole located at about 47°N and 170°E, without drifting perceptibly northward. Motions apparently accelerated during the Westphalian. The central land mass of Laurussia continued to straddle 20°N latitude. Progressive uplift of the Hercynian Orogen, subparalleling the equator, combined with the Westphalian establishment of paralic conditions in the Appalachian and Variscan foreland basins, and continental conditions in northwest Africa, provided a coherent landbridge between Gondwana and Laurussia. Already during the late Namurian the circum-equatorial current through North Africa became blocked and with this the dispersal of the formerly cosmopolitan shallow marine, warm water faunas became interrupted; this caused the progressive development of a new faunal provincialism (Ross and Ross, 1985). At the same time the phytobiogeographic distinction between the former North Gondwana and European-Acadian provinces disappeared (Rowley et al., 1985).

Progressive glaciation of southern Gondwana was accompanied by cyclical sea-level fluctuations. However, major regressions and transgressions on the shelves of Laurussia can be related, to a large extent, to intra-plate deformations.

Overall, Namurian and Westphalian times were a period of increased tectonic instability of cratonic areas flanking the Hercynian fold belt. In the United States, upwarping of the Transcontinental Arch during the Namurian caused a major regression that was followed by a regional transgression during the Westphalian phases of foreland faulting. Uplift of the West Canadian Platform at the end of the Visean caused a substantial enlargement of land areas that was, however, compensated by the rapid expansion of the Arctic carbonate shelf. The carbonate-dominated Moscow platform underwent only minor modifications in its outline during the Namurian and Westphalian and remained, overall, tectonically quiescent.

Evolution of the Norwegian-Greenland Sea Rift and the northeast Greenland-Svalbard oblique-slip zone paved the way for the southward transgression of the Arctic and Barents Shelf seas and for their connection during the Westphalian. Geodynamic processes governing the development of the Norwegian Greenland Sea Rift remain uncertain. However, the absence of volcanic events (apart from the Kontozero rift), precludes an active mantle plume model. On the other hand, tensional reactivation of fault systems, that formed part of the Devonian-early Carboniferous Arctic-North Atlantic megashear, in response to the build-up far-field compressional stresses emanating from the Variscan fold belt, may be the cause for the initiation of rifting activity. Significantly, the strike of the Norwegian-Greenland Sea Rift appears to run sub-parallel to the principal horizontal compressional stress axes in the Variscan foreland. Moreover, the onset of rifting activity in the Norwegian Greenland Sea coincides with the closure of the Rhenohercynian back-arc basin and the development of the Variscan foreland basin. The late Carboniferous configuration of the Norwegian-Greenland Sea rift system suggests a northward increase in divergence between Greenland and Fennoscandia, that was compensated by a dextral translation between the Arctic Shelf and the Canadian Arctic Archipelago. At the same time, the Devonian and Carboniferous grabens of the British Isles and the Canadian Maritime Provinces, which trend obliquely to the inferred foreland compressional stresses, became partially inverted during the late Westphalian phases of the Variscan Orogeny.

Overall, the late Namurian and Westphalian Hercynian collision between the Gondwana and Laurussia megacontinents resulted in important changes in the palaeotectonic and palaeogeographic setting of cratonic Laurussia and northern Gondwana. This was paralleled by important changes in faunal and floral provincialities.

7. Stephanian-Autunian

The latest Carboniferous-earliest Permian configuration of Pangea is summarised in Plate 2c. An overview of the palaeogeographic-palaeotectonic setting of the Laurussian subcontinent is presented in Plate 10.

7.1. Plate Boundaries and Continent Assembly

During the Stephanian and Autunian (Missourian-early Leonardian), Gondwana continued to converge with Laurussia. Increased coupling between these two megacontinents is evident by their joint northward drift of some 8°.

While crustal shortening in the Marathon-Ouachita-Appalachian Orogen continued into early Permian times, the Variscan fold belt became transsected by a system of post-orogenic, dextral wrench faults that caused its collapse and partial destruction. This suggests that from the latest Westphalian to the early Stephanian the direction of convergence between Gondwana and Laurussia changed from essentially north-south during the Variscan Orogeny, to northwest-southeast during the terminal phases of Alleghanian Orogeny. The wrench fault system, that transsected the Variscan fold belt remained active until the early Permian consolidation of the Marathon-Ouachita-Appalachian fold belt. The post-Variscan shear fault system, that transects Western and Central Europe, represents a diffuse plate boundary between the Africa-Proto Tethys Plate and Laurussia. It linked the southern end of the Uralian Orogen with the northern end of the Appalachian Orogen (Arthaud and Matte, 1977; Ziegler, 1982, 1988).

The Permo-Carboniferous evolution of the eastern margin of Laurussia was dominated by the closure of the Sakmarian Ocean and the collision of Kazakhstan with the passive margin of the Moscow Platform. At the same time Siberia continued to converge with Kazakhstan, as reflected by orogenic activity along the Kuzbass-Tom-Kolyva suture (Zonenshain et al., 1987b). By Autunian time the collision zone between Kazakhstan and the eastern margin of Laurussia had propagated northward into the Timan-Pechora area (Pay Kohy). This was paralleled by the closure of oceanic domains separating the Barents Shelf and Siberia. To the north, the Ural-Novaya Zemlya subduction zone probably linked up with the intra-oceanic Arctic arc-trench system, that presumably evolved in response to the northward drift of Laurussia (Zonenshain et al., 1978b; Zonenshain, personal communication). This arc-trench system was probably connected to the southwest with the intra-oceanic Sonoma arc, located to the west of the North American Craton. Based on limited data, it appears that the Cordilleran miogeocline was tectonically quiescent during the latest Carboniferous and early Permian. Its carbonate-dominated shelves faced the Havallah back-arc basin, which was limited to the west by the active Sonoma arc (Burchfiel and Davis, 1975; Frazier and Schwimmer, 1987).

The Norwegian-Greenland Sea rift and the trans-Arctic oblique-slip zone remained active during the Stephanian-Autunian. Stresses related to the post-Variscan wrench fault system presumably facilitated the southward propagation of the Norwegian-Greenland Sea Rift.

With the early Permian consolidation of the Marathon-Ouachita-Appalachian-Mauretanides fold belt the Hercynian suturing process of Laurussia and Gondwana was completed.

7.2. Post-Variscan Wrench Faulting

The change in plate motions, that underlies the latest Westphalian termination of orogenic activity in the Variscan fold belt and the locking of its subduction system, caused a major dextral translation between Africa and Europe. In the domain of the Variscan fold belt, this translation resulted in the development of a complex set of conjugate shear faults, the subsidence of transtensional and pull-apart basins, and transpressional deformations that gave rise to the formation of tectonically silled basins. Activity along this fault system, that extended into the Variscan foreland, was associated with deep crustal fracturing and widespread volcanism of a highly variable chemism (Francis, 1988).

The dimensions of the Stephanian-Autunian dextral translation between Africa and Europe are difficult to quantify, but should be equal to the amount of crustal shortening achieved during the late phases of the Alleghanian orogeny in the Appalachian-Mauretanides. In the model presented here, it is assumed that these displacements amounted to at least 200 km (see Secor et al., 1986).

It is uncertain whether, and to what degree, back-arc extension, related to the decay of the north-plunging Variscan subduction system,

also played a role in the Stephanian and Autunian tectonic evolution of Western and Central Europe (Lorenz and Nicholls, 1984; Ziegler, 1988).

Details of the post-Variscan fault system can only be mapped in the extra-Alpine outcropping Variscan massifs and in the subsurface of sedimentary basins that have been explored for hydrocarbons. Fault patterns shown in Plate 10 for the Alpine and Mediterranean areas are schematic (Arthaud and Matte, 1977; Ziegler, 1982).

In the Alpine foreland, main elements of the Permo-Carboniferous fault system are the Tornquist-Teisseyre and the Bay of Biscay fracture zones. Dextral displacement along the Tornquist-Teisseyre line induced the development of the highly volcanic Oslo Rift and of major volcanic fields in northern Germany and Poland that also appear to be associated with pull-apart features. Displacement along the Bay of Biscay fracture zone were accompanied by transpressional deformations in Cantabria and the subsidence of the Western Approaches and Porcupine troughs, and possibly by crustal extension in the Rockall-Faeroe Rift. It is speculated that tensional stresses, developing at the northwestern termination of the Bay of Biscay fracture zone and of subsidiary fault systems crossing the British Isles, may have facilitated the southward propagation of the Norwegian-Greenland Sea Rift (Ziegler, 1982, 1988; see also Tate and Dobson, 1988a).

The Gibraltar and Agadir fracture zones, although less well documented, are thought to mark the northern termination of the Alleghanian-Appalachian Orogen. Activity along these fault systems was probably associated with the late Autunian partial inversion of the Permo-Carboniferous Magdalen, Sidney and possibly also the St. Anthony basins in the Canadian maritime provinces

(Maritime Disturbance; Howie and Barss, 1975).

Areas located between these four principal fracture zones became dissected by a complex set of conjugate shear faults. These became reactivated repeatedly during the Mesozoic rifting phases that preceded the back-up of Pangea.

In Western and Central Europe the Stephanian-Autunian tectonism induced a regional regression and deep truncation of sedimentary basins in the Variscan foreland. Syntectonic deposits consist of continental, often coarse, coal-bearing series that grade northward into barren red beds (Ziegler, 1982, 1988).

7.3. Alleghanian Consolidation of Appalachian Orogen

In the Marathon-Ouachita-Appalachian-Mauretanides Orogen, crustal shortening persisted on a grandiose scale during the latest Carboniferous and early Permian (Hatcher, 1981; Kulander and Dean, 1986; Wickham et al., 1976; Blythe et al., 1988). In the Appalachian foreland basin, and in the Arkoma and Fort Worth basins (Ouachita foredeep), cratonward migration of the basin axis was accompanied by thin-skinned thrusting and the accumulation of molasse-type clastics. However, in the Marathon foredeep, consisting of the Val Verde and Marfa Basin, deepwater conditions persisted through latest Carboniferous and early Permian times (Ross and Ross, 1985a). Crustal shortening is thought to have persisted in the Appalachian Basin into the Wolfcampian, in the Ouachita foredeep until the transition from the Virgilian to the Wolfcampian, and in the Marathon foredeep until late Wolfcampian time (Frazier and Schwimmer, 1987; Wuellner et al., 1986). As indicated earlier, the timing of deformation of the Mauretanides is poorly constrained, due to the

subsequent destruction of synorogenic deposits. In the Canadian maritime provinces the Maritime Disturbance can be dated as intra-Autunian. The termination of the Alleghanian diastrophism thus post-dates the last orogenic pulses of the Variscan fold belt by some 30 Ma.

The northernmost, truely orogenic Alleghanian deformations, involving significant crustal shortening, occurred in the New England States and in New Brunswick. Here polyphase deformations, associated with thrust faulting, were accompanied by metamorphism and granitic intrusions ranging in age from 380 to 260 Ma (Dallmeyer, 1982; Dallmeyer and Keppie, 1987; Mosher and Rast, 1984; Rast, 1983; Reynolds et al., 1981; Skehan, 1983; Wintsch and Sutter, 1986; Nance, 1987). These deformations were coupled with important dextral movements along the Minas Geofracture, a palaeomagnetically documented 20° rotation of the Meguma Terrane (Scotese et al., 1984; Spariosu et al., 1984; Keppie, 1985) and the partial inversion of the Magdalen and Sidney Basin during the so-called Maritime Disturbance (Howie and Barss, 1975).

For the Appalachians, COCORP deep reflection seismic data (Cook, 1983; Cook et al., 1979, 1981) indicate that Alleghanian crustal shortening amounts to at least 175 km (Secor et al., 1986). For the entire Appalachian-Mauretanides megasuture, the latest Carboniferous-early Permian crustal shortening may be as much as 300 km. Alleghanian consolidation of this fold belt was accompanied by the widespread intrusion of S-type granites, ranging in age from 325 to 265 Ma (Hatcher, 1987). During the late phases of the Alleghanian diastrophism collisional overthickening of the lithosphere was accompanied by thrust propagation deep into the North American foreland. Sedimentation in the Appalachian foreland basin continued at least during the early phases of its deformation (Secor et al., 1986; Hatcher, 1987; Beaumont et al., 1987). At the same time the Cincinnati and Adirondack arches became accentuated by intra-plate compressional stresses. Uplift of the Adirondacks induced deep truncation of the northern most parts of the Appalachian foreland basin. On the other hand, the Nashville Dome became apparently deeply buried under molasse-type clastics (Beaumont et al., 1987).

Also, the Mauretanides salient apparently underwent a last phase of deformation during the latest Carboniferous-earliest Permian (Lécorché and Clauer, 1983; Vauchez et al., 1987), as evident from radiometric age determinations (Dallmeyer and Villeneuve, 1987). This was coupled with a further uplift of the Reguibat Arch, separating the Taoundeni and Tinduf basins. The late phases of the Alleghanian orogeny apparently were associated with dextral strike slip movements in the internal parts of the Appalachian fold belt (Vauchez et al., 1987; Wintsch and Sutter, 1986).

The Alleghanian consolidation of the Marathon-Ouachita segment of the Hercynian suture between North America and Gondwana probably involved an equal, if not even greater amount of crustal shortening than was achieved in its Appalachian-Mauretanides segment. Thin skinned thrusting and northward migration of the Marathon-Ouachita fore-deep basin axes was paralleled by continued foreland deformations in the area of the ancestral Rocky Mountains. This is reflected by the accentuation of, for instance, the Uncompahgre, Amarillo-Wichita and Pedernal uplifts and the differential subsidence of the the Central Colorado Trough, the Delaware, Midland and Anadarko basins (McKee and Crosby, 1975; Ross and Ross, 1985a, 1986; Wuellner et al., 1986; Kluth, 1986; Frazier and Schwimmer, 1987).

The westward continuation of the Ouachita-

Marathon fold belt into Mexico and its possible link-up with the Sonoma arc-trench system and the Andean Hercynides is unclear (Shurbert and Cebull, 1987; Handshy and Dyer, 1987; Handshy et al., 1987; Stevens and Stone, 1988; Armin, 1987).

In view of the limited exposure of the Hercynian basement complex in the area of the Gulf of Mexico and its Mesozoic and Cenozoic disruption and overprinting, reconstruction of this part of the Pangean Hercynian megasuture has to contend with major uncertainties (Pindell and Dewey, 1982; Pindell, 1985). The actual suture between Gondwana and North America is thought to sub-parallel the coastline of Senegal and to cross northern Florida, where it is interpreted to coincide with the Brunswick magnetic anomaly (Tauvers and Muehlberger, 1987; Dallmeyer, 1987; Opdyke et al., 1987; Vauchez et al., 1987). Its westward continuation is uncertain.

In northern Venezuela, carbonate-clastic sedimentation persisted into the early Permian, by which time the area became deformed. From surface and sub-surface data it is assumed that the Hercynian deformation front subparallels the Meride Andes, but is concealed beneath the Mesozoic and Tertiary Barinas-Apure and East Venezuelan foreland basins. Northern Venezuela is characterised by widespread late Carboniferous and Permian granitic intrusives (Juana et al., 1980). A typical Permo-Carboniferous foreland basin seems to be absent. From the distribution of granites it appears that the external parts of the Venezuelan Hercynian fold belt have been destroyed by erosion during subsequent tectonic events.

7.4. Uralian Orogenic Systems

Continued convergence of Siberia and Kazakhstan with each other and the eastern margin of

Laurussia involved the northward propagation of the collision front between Kazakhstan and the Moscow Platform and the gradual closure of oceanic domains separating the Siberian Craton and the Barents Shelf. This was paralleled by the development of the Ural and Novaya Zemlya foreland basins (Zonenshain et al., 1984; Artyushkov and Baer, 1983; Ronov et al., 1984; Sobolev, 1982).

Subsidence of the Uralian foreland basin, in response to tectonic loading of the Fennosarmatian Shelf by the advancing nappe systems, was accompanied by the accumulation of synorogenic flysch in its proximal parts and the progressive drowning of its distal carbonate platform. In the southern parts of the Ural foredeep, influx of clastics from the rising orogen commenced during the Bashkirian-Moscovian (early Westphalian), whereas in its northern parts, the Timan-Pechora Basin, earliest flysch series are dated as Stephanian. In the foreland of Novaya Zemlya, the Barents Shelf carbonate platform subsided rapidly during the Asselian and by Artinskian time clastic influx from eastern sources is evident (Heafford, 1988). This suggests that by earliest Permian time the collision front between Kazakhstan-Siberia and the eastern margin of Laurussia had propagated at least as far north as Novaya Zemlya. During the Autunian, compressional stresses exerted on the north Uralian foreland were of sufficient magnitudes to initiate the inversion of the Devonian-early Carboniferous Pechora-Kolva grabens (Matviyevskaya et al., 1986).

In the area of Taimyr Peninsula, late Carboniferous and early Permian clastics, contained in the South Taimyr Trough, were derived both from a southern and a northern source (Vinogradov, 1969; Ulmishek, 1984). The presence of a northern clastic source area, that later may have become overriden by the Kara Massif during the late Permian and Trias-

sic final deformation of the Taimyr fold belt (Churkin et al., 1981; Ulmishek, 1984; Pogrebitsky, 1982), suggests that also in the northernmost parts of the Uralian system continent-to-continent collision had occurred or was imminent, during the Permo-Carboniferous.

On the Siberian Platform, late Carboniferous and early Permian sediments are developed in a continental, coal-bearing molasse facies (Vinogradov, 1969; Ronov et al., 1984).

Late Carboniferous and early Permian granitoides are widespread both in the Ural and in the Kusbass-Tom-Kolyva fold belt (Vinogradov, 1969).

7.5. Norwegian-Greenland Sea Rift and Arctic Shelves

The graben systems of the Norwegian-Greenland Sea area and the western Barents Shelf remained active during the Stephanian-Autunian. In Central East Greenland continental red beds accumulated in half grabens, whereby the onset of sedimentation became increasing younger towards the south. An important rifting pulse, giving rise to a regional unconformity, is evident at the transition from the Stephanian to the Autunian (Surlyk et al., 1986). Southward propagation of the Norwegian-Greenland Sea rift into the Faeroe and Rockall trough cannot be substantiated, but is inferred from the late Permian facies patterns (Plate 12). On the West-Shetland shelf, poorly dated Permo-Triassic red beds, accumulating in rapidly subsiding half-grabens, may extend into the Stephanian-Autunian (Ziegler, 1988).

Continued subsidence of the Tromsø-Byørnøya and Nordkap grabens on the Western Barents Shelf is suggested by the development of cherty-shaly, starved basin conditions, whereas carbonates continued to be deposited on the more stable platforms. On Svalbard, the latest Carboniferous and early Permian series consist of partly reefal carbonates, grading upwards into evaporites; their deposition was fault-controlled (Steel and Worsley, 1984; Heafford, 1988). Similarly, there is evidence for syn-depositional faulting in the northeast Greenland Wandel Sea Basin (Håkansson and Stemmerik, 1984). This reflects continued tectonic activity along the trans-Arctic oblique-slip zone, albeit at a lower level than during the Namurian and Westphalian.

In the Canadian Arctic Archipelago, rifting gradually abated and the Sverdrup Basin began to assume the geometry of a thermal sag basin during the latest Carboniferous and early Permian (Sweeney, 1977; Stephenson et al., 1987). During the Stephanian-Autunian clastic input into the Sverdrup Basin was at a low level. Carbonate and carbonate-clastic banks, occupying its margins, were offset by shales that reflect starved basin conditions (Rayer, 1981). Early Permian volcanics, occurring in northern Ellesmere Island (Wynne et al., 1983), testify, however, to continued activity along the trans-Arctic oblique slip zone.

As during the Westphalian, the Arctic and Barents shelves were occupied during the latest Carboniferous and early Permian by carbonate and mixed carbonate-clastic platforms. These may have extended across the Lomonosov High (Heafford, 1988; Ulmishek, 1985; Fuijta and Cook, 1986). Intensified uplift of the southern parts of Chukotka is reflected by a clastic apron on its northern parts. The mechanism of this uplift is unknown, but could be related to the subduction process along the northern margin of Chukotka and the evolution of the Arctic island arc (Zonenshain et al., 1987b); the latter probably linked the Ural-Taimyr subduction system with the Sonoma

arc paralleling the western margin of the North American Craton.

7.6. Platform Areas

As a consequence of the Stephanian and Autunian northward drift of Pangea, the area of the central Laurussian land mass straddled 20°N latitude and continued to be dominated by an arid climate. The Arctic-Barents Sea carbonate shelf was located between 40° and 45°N latitude. Despite the increasing glaciation of southern Gondwana (Hambrey and Harland, 1981), causing a cyclical eustatic lowering of global sea-levels, the area occupied by shelf seas fringing the Laurussian land mass did not change substantially. Regressive conditions on the Moscow Platform, combined with its progressive tectonic isolation in conjunction with the Uralian Orogeny, led to the establishment of evaporitic conditions.

Regional uplift of the area of Western and Central Europe can be related to post-Variscan wrench faulting, involving broad scale, as well as local, lithospheric deformations and the introduction of important thermal anomalies.

The West Canadian Platform and the Labrador-Greenland High formed a coherent, stable landmass, from which only limited clastics were shed into the adjacent basins. The area of the United States was occupied by a tectonically metastable marine shelf, which received clastics from the rising Appalachian-Ouachita Orogen, from the active Uncompaghre and Frontrange uplifts, as well as from the West Canadian Platform. The western margin of this wide shelf was occupied by carbonate platforms, facing the oceanic "Havallah" back-arc basin. In comparison, the Canadian Cordilleran miogeoclinal basin was presumably relatively narrow; its palaeogeographic framework is poorly constrained and is shown schematically only on Plate 10 (Cook and Bally, 1975; McKee and Crosby, 1975; Frazier and Schwimmer, 1987; Gabrielse, 1967; Ziegler, 1969).

Along the northern margin of Gondwana, carbonate shelves were maintained during the Stephanian and Autunian on the Northeast African-Arabian platform and in the foreland of the Colombian Andes; the latter may have extended into Venezuela. These shelf seas were separated by a coherent land mass, in the development of which uplift of major arches (Guayana Shield, Reguibat, Ougarta, El Bjot uplifts) played an important role. Within this land mass, accumulation of continental clastics may have continued in tectonically silled basins into the early Permian. By this time regional post-orogenic uplift of the Appalachian Orogen and its foreland basins induced deep truncation of their sedimentary fill (Beaumont et al., 1987).

8. Late Early And Late Permian

The late early Permian (Rotliegend, Saxonian, late Artinskian-Kungurian) and the late Permian (Zechstein, Thuringian, Kazanian-Tatarian) palaeogeographic-palaeotectonic configuration of the northern parts of Pangea are summarised in Plates 11 and 12. An overview of the Pangea continent assembly and its palaeolatitudinal position is provided by Plate 2e and f.

8.1. Plate Boundaries and Continent Assembly

With the Alleghanian consolidation of the Marathon-Ouachita-Appalachian fold belt the Hercynian suturing process of Gondwana and Laurussia was completed. Together these megacontinents formed the core of the Pangea supercontinent. With this Laurussia had lost its identity.

Orogenic activity continued, however, in the Uralian system during the late Permian-early Triassic, by which time Siberia and Kazakhstan were accreted to the eastern margin of the Laurussian sub-continent. Moreover, Gondwana-derived continental terraines, now forming parts of China, continued to converge and collide, during the late Permian and Triassic with Kazakhstan and Siberia (A.M. Ziegler, 1981; Parrish et al., 1986).

The late Hercynian Uralian subduction system probably was linked with that of the Cordillera via the Arctic arc-trench system that paralleled the northern margin of the Laurussian sub-continent (Zonenshain et al., 1987b). In the North American Cordillera, orogenic activity accelerated during the late Permian and culminated in the Sonoma Orogeny at the transition of the Permian to the Triassic (Burchfiel and Davis, 1972, 1975; Frazier and Schwimmer, 1987; Walker, 1988). Similarly, there is evidence in the Andean system for a major orogenic event during the late Permian and earliest Triassic (Carlier et al., 1982; Mesner and Woodridge, 1964). This also coincides with the development of the South African Cape fold belt foreland basin (Hälbich, 1983) and continued orogenic activity along the Pacific margin of Antarctica and Australia (St. John, 1986).

The overall impression gained is, that locking of the Hercynian megasuture between Gondwana and Laurussia, as a consequence of overthickening of the lithosphere, combined with continued activity along the Pacific sea-floor spreading axes, caused the reactivation of subduction zones along the western and southern margin of Pangea. During the late early and late Permian this was coupled with a counterclockwise rotation of Pangea, amounting to about 20°. The rotation pole for this motion was located in the western Gulf of Mexico. As a consequence of this rotation, northern Siberia moved into a circum-polar position. At the same time Antarctica moved out of a location centred on the South Pole. By late Permian time ice caps were present in northern Siberia and in Antarctica (Ustritsky, 1973; Hambrey and Harland, 1981; Caputo and Crowell, 1985). The transfer of subduction processes from the central Pangean Hercynian suture to the southern periphery of Pangea reflects a fundamental re-organization of plate boundaries.

During the Permian, the Norwegian Greenland Sea rift remained active and its subsidence paved the way for the late Permian southward transgression of the Arctic Sea. On the shelves flanking the western parts of the Tethys embayment, rifting activity commenced, probably in conjunction with the slow development of the Mesozoic Tethys sea floor spreading axis. Late Permian opening of an oceanic basin in the Black Sea may be related back-arc extension, induced by the decay of the Variscan subduction zone. Permian acceleration of rifting activity in the western Tethys domain, in the Norwegian Greenland Sea and the inception of the Gondwana rifts (East Africa, India, western Australia) heralds the fundamental plate re-organization that underlies the Mesozoic break-up of Pangea (Ziegler, 1988; Dewey, 1988).

8.2. Uralian and Cordilleran Orogenic Belts

The Ural-Novaya Zemlya-Taimyr orogenic system remained active through Permian time. Last orogenic pulses occurred at the transition to the Triassic. The Novaya Zemlya, and particularly the Taimyr fold belt became re-activated during the Triassic-earliest Jurassic Indosinian orogenic pulse, that is of major importance in South East Asia.

In the southern Ural foreland basin, early Permian flysch series, overlain by Kungarian molasse-type clastics, prograded over the earlier carbonate shelf. This clastic wedge was separated from the carbonate dominated Moscow Platform by a deeper water, sediment starved trough, that became infilled with thick halites during the late Kungurian. During the terminal phases of the Uralian Orogeny, late Permian and early Triassic continental clastics accumulated in this foredeep. The structural style of the external zones of the Urals is characterised by thin-skinned thrust sheets (Slezinger and Jansin, 1979; Nalivkin, 1982; Artyushkov and Baer, 1983; Dymkin et al., 1984; Kazantseva and Kamaletdinov, 1986).

In the northern Urals (Pay Kohy), flysch sedimentation gave way to the accumulation of shallow marine, molasse-type clastics at the transition from the Asselian to the Sakmarian. During the Kungurian to Tatarian, 3 to 4 km thick continental and paralic, coal-bearing clastics accumulated in the Pechora foredeep. At the same time inversion of the Pechora-Kolva rifts and uplift of the Timan Range continued. Deposition of coal measures ceased at the transition from the Permian to the Triassic and gave way to the accumulation of red beds. During the early Triassic large clastic fans prograded westward over the carbonate platforms of the Pechora and Barents shelves; to the west these were offset by basinal, partly organic shales. The Uralian orogeny terminated during the early Triassic. However, during the late Triassic-early Jurassic the northern parts of the Urals became re-activated again (Churkin and Trexler, 1981; Nalivkin, 1982; Ulmishek, 1982; Vinogradov, 1969; Matviyevskaya et al., 1986; Heafford, 1988).

The early Permian deepwater foreland basin of Novaya Zemlya became filled by 2 to 3 km thick Kazanian and Ufimian flysch-type clastics. During the late Ufimian shallow marine deltaic conditions prevailed. Folded Permian series are overlain by thick, less intensely folded Triassic clastics, which became deformed during the late Triassic-early Jurassic. Consolidation of the Novaya Zemlya fold belt was punctuated by the intrusion of late Permian-early Triassic and late Triassic-early

Jurassic granitoides (Churkin et al., 1981; Romanovitch, 1982; Ulmishek, 1985; Gramberg et al., 1986; Heafford, 1988).

In the South Taimyr trough, late early and late Permian paralic, coal-bearing clastics, derived from northern and southern sources, reach thicknesses of the order of 3 to 5 km. This testifies to the continued uplift of a northern clastic source area, probably corresponding to the Kara Massif; this massif was thrusted southward over the South Taimyr trough during the early Triassic. A second phase of compressional deformation occurred at the transition from the Triassic to the Jurassic. During the late Permian and early Triassic extensive plateau basalts, 1 to 3 km thick, were extruded in the South Taimyr trough. The Kara Massif was affected by metamorphism and was intruded by granites, yielding ages of 250-220 Ma. The South Taimyr trough can be considered as a retro-arc foreland basin that is superimposed on the margin of the Siberian Craton. Whether a conjugate fore-arc foreland basin, forming the continuation of the Novaya Zemlya foredeep, parallels the northern margin of the Kara Massif is unknown (Ulmishek, 1985; Churkin et al., 1981; Pogrebitsky, 1982; Vinogradov, 1969; Gramberg et al., 1986).

Unlike the Urals and the Novaya Zemlya fold belts, the Taimyr fold belt lacks ophiolites and must therefore be considered as a cryptic suture. As such, it forms an integral part of the Ural-Novaya Zemlya orogenic system. This megasuture, along which Laurussia and Siberia/Kazakhstan were welded together during the late Permian, became reactivated during the Indosinian orogenic pulse at the transition from the Triassic to the Jurassic.

During the late Permian and Triassic the area of the future Mesozoic West Siberian Basin was affected by tensional tectonics that were accompanied by the outpouring of extensive plateau basalts (see also South Taimyr trough above). This is interpreted as being the expression of back-arc extension that was induced by the gradual decay of the east-dipping Ural-Taimyr subduction zone (Rudkevich, 1976; Aleinikov et al., 1980; Chablinskaia et al., 1982).

Plates 11 and 12 suggest that the Taimyr subduction system was connected via the hypothetical Arctic island arc with the Cordilleran arc-trench system (see also Zonenshain et al., 1987b). Late early and late Permian counter-clockwise rotation of Laurussia and its convergence with the Arctic sector of the Pacific Plate was coupled with increased tectonic activity along the northern continental margin of Laurussia, possibly involving the development of a subduction zone. This is reflected by the progressive uplift of the southern parts of Chukotka and the uparching of the Lomonosov High (Grumant High; Vinogradov, 1969; Ulmishek, 1985).

In the North American Cordilleran system, tectonic activity intensified during the late Permian and culminated at the transition from the Permian to the Triassic in the Sonoma Orogeny. This reflects an increase in the convergence rate between the oceanic Pacific Plate and the North American Craton. During the Sonoma Orogeny, the Havallah back-arc basin was closed and the Sonoma arc-terrane became accreted to the margin of the Cordilleran Miogeocline (Burchfiel and Davis, 1972, 1975; Frazier and Schwimmer, 1987). In the latter this was paralleled by increased clastic influx from the Cassiar-Yukon-Fairbanks Platform and the development of starved basin conditions in its distal parts, where silicious shales and cherts were deposited. In the Canadian sector of the Cordilleran Miogeocline the late Permian corresponds to a transgressive phase (Ziegler, 1969). On the the Alaska North Slope, intra-Artinskian uplift of its northern parts gave rise to a regional regression and the termination of carbonate deposition. Late Artinskian to Tatarian transgressive series are developed in a deltaic to shallow marine clastic facies. To the south this series grades into silicious, basinal shales (Hubbard et al., 1987).

8.3. Hercynian Megasuture

The early phases of the post-orogenic development of the Marathon-Ouachita-Appalachian-Mauretanides fold belt is difficult to assess for want of a corresponding stratigraphic record.

Post-orogenic uplift and progressive erosional unroofing of the Appalachian-Mauretanides fold belt is reflected by retrograde metamorphism. Quantitative analyses in the New England States suggest that this area became uplifted by some 25 km during the late Permian and Triassic (Wintsch and Sutter, 1986). At the same time the area of the Appalachian foreland basin became uplifted and subjected to erosion (Beaumont et al., 1987). Whether this late Permian regional uplift was accompanied by tensional tectonics is uncertain, since the age of the earliest deposits in the Triassic grabens of the Appalachian domain is largely unknown (see Manspeizer, 1988).

Overall, the Central Atlantic domain corresponded, during the late Permian and early Triassic, to a coherent highland that was apparently tectonically quiescent, apart from isostatic uplift in response to erosional unroofing of its axial Hercynian fold belt. A similar history could be envisaged for the late Permian evolution of the Hercynian fold belt occupying the area of the Gulf of Mexico.

Post-orogenic evolution of the Permian basins in the foreland of the Ouachita-Marathon fold belt was characterised by tectonic quiescence, the gradual infilling of their remanent basin

topography by carbonates and evaporites and a progressive restriction of the area of sedimentation (Cook and Bally, 1975; Frazier and Schwimmer, 1987). Within the domain of the Gulf of Mexico, sedimentation resumed during the late Triassic in rift/wrench induced basins that evolved during the early phases of the Pangea break-up (Walper, 1981; Pindell, 1985).

In contrast to the above, the area of the Variscan fold belt provides a record of late Permian rifting that can be related to the gradual activation of the Tethys sea floor spreading axis that governed the Mesozoic break-up of the Pangea (Ziegler, 1988).

Rapid degradation of the Variscan fold belt in the Mediterranean area was coupled with the establishment of late Permian carbonate platforms fringing the wedge-shaped Tethys embayment (Kamen-Kay, 1976; Argyriadis et al., 1980; Tollman, 1984). However, in view of palinspastic uncertainties, the reconstructions of the Mediterranean area, as shown in Plates 11 and 12, must be considered as conceptual.

Development of late Permian graben systems, punctuated by rift volcanism, is indicated for the Balkan, Hellenic, Dinarid and South Alpine domains. A major feature is the Pindos-Subpelagonia Trough (SPT), that isolated the Pelagonia-Golija block from the Italo-Dinarid platform (Pamic, 1984; Roddick et al., 1984). In the Black Sea area, Permian back-arc rifting, probably related to the decay of the Variscan subduction system, culminated in late Permian-early Triassic seafloor spreading and the separation of the Variscan deformed Cimmeria Terrane from the southern margin of Fennosarmatia (Adamia et al., 1981; Khain, 1984). From the Black Sea area rifting propagated apparently northwestward into Poland.

In areas flanking the Tethys shelves, tectonic activity abated rapidly with termination of wrench faulting during the latest Autunian. Within the domain of the Variscan fold belt continental series accumulated in intra-montane depressions. The gradual development of a regional tensional regime is reflected by the late Permian and particularly early Triassic re-activation of Permo-Carboniferous fault systems and the development of new multidirectional grabens (Ziegler, 1982; 1988).

8.4. Norwegian-Greenland Sea Rift and Arctic Shelf

In the Norwegian-Greenland Sea area an important rifting pulse, accompanied by dyke intrusions, occurred during the Kungurian. It resulted in further tilting of major basement involving rotational fault blocks and uplift of their leading edges. This rifting phase preceded the transgression of the Late Permian seas that gave rise to the deposition of carbonates and evaporites in central East Greenland and in the Mid-Norway shelf basin. A further rifting pulse marked the Permian-Triassic boundary (Vischer, 1943; Haller, 1971; Stemmerik and Sørensen, 1980; Surlyk et al., 1986). By late Permian time, rifting had propagated into the Faeroe and Rockall Trough, as evident by the occurrence of evaporitic series in the West Shetland Basin and on the north Irish Shelf (Tate and Dobson, 1989b) and possibly also into the northernmost North Sea. Through these avenues the Arctic Sea ingressed into the Northern and Southern Permian basins of Western and Central Europe, where it gave rise to the accumulation of the highly cyclical Zechstein carbonate, sulphate and halite series (Ziegler, 1982, 1988; Ziegler and van Hoorn, 1989).

The Northern and Southern Permian Basin began to subside differentially, during the Saxonian, in areas that had been affected by intense Stephanian-Autunian wrench and pull-apart tectonics and volcanism. The geometry of the Southern Permian Basin, which is well constrained by drilling and seismic data, indicates that it evolved as a broad, saucer-shaped downwarp, presumably in response to the decay of a thermal anomaly that was introduced during the post-Variscan phase of wrench faulting. A similar mechanism may also apply for the development of the Northern Permian Basin (Ziegler, 1982, 1988; Ziegler and van Hoorn, 1989).

Only in the eastern parts of the Southern Permian Basin exists evidence for fault-controlled subsidence of the Polish Trough; this graben corresponds to the northwestern part of the Black Sea-Dobrugea rift system.

Thick, continental clastics accumulated in the Northern and Southern Permian Basin during the Rotliegend. The central parts of the latter were occupied by an evaporitic playa lake. The transgression of the late Permian Zechstein Sea was apparently catastrophic, suggesting that these basins had subsided below the global sea-level prior to their flooding. During early Zechstein times a temporary connection was established between the Arctic and the Tethys Seas via the Polish-Dobrugea graben (Ziegler, 1982). Zechstein facies patterns along the southern margin of the Southern Permian Basin are indicative for the gradual development of a tensional regime that heralded the onset of the Triassic rifting phase. There is, however, clear evidence that the Viking and Central Graben of the North Sea had not yet started to subside during the late Permian (Ziegler and van Hoorn, 1989).

In the area of the western Barents Sea, Svalbard and northeast Greenland, rift and wrench tectonics continued during the late Permian and accelerated sharply at the transition from the Permian to the Triassic. During the late Permian carbonate platforms were maintained on stable blocks, whereas siliceous series accumulated in differentially subsiding grabens and wrench induced troughs. Uplift of the Lomonosov High is reflected by clastics shed onto the carbonate platform of the eastern Svalbard Archipelago. Also in Northeastern Greenland, increased tectonic instability during the late Permian is reflected by the deposition of shallow-marine, conglomeratic sands (Håkansson and Stemmerik, 1984; Steel and Worsley, 1984; Heafford, 1988; Ziegler, 1988).

Crustal extension across the Norwegian-Greenland Sea Rift was compensated by continued tectonic activity along the trans-Arctic oblique-slip zone. This is evident by the Melvillian disturbances that affected the Sverdrup Basin during the late Permian. At the same time clastic sources became uplifted in northern Ellesmere Island and in the New Siberian Islands. Moreover, the late early Permian uplift of an arch, paralleling the coast of the Alaska North Slope, may also be related to movements along the trans-Arctic fracture zone.

On the Arctic and Barents shelves carbonate deposition persisted during the late Permian but terminated abruptly at the transition from the Permian to the Triassic (Fujita and Cook, 1986; Heafford, 1988). This was probably the combined effect of northward drift of Pangea and increased tectonic activity along the northern margin of the Laurussian sub-continent.

9. Discussion

Plate reconstructions underlying our analysis of the evolution of Laurussia are tentative and subject to revision as new palaeomagnetic and palaeoclimatological data become available.

Nevertheless, it appears that after the latest Silurian suturing of Laurentia-Greenland and Fennosarmatia along the Arctic-North Atlantic Caledonides, the Devonian evolution of Laurussia was governed by the convergence of the Proto Atlantic-Proto Tethys, Ural, Pacific and Arctic plates. These plate motions were coupled with the accretion of Avalonia, the Aquitaine-Cantabrian and the ill-defined intra-Alpine terranes to the southern margin of Laurussia during the middle Devonian Acadian-Ligerian Orogeny and the suturing of the Arctic craton to its northern margin during the early Mississippian Ellesmerian diastrophism. Moreover, at the transition from the late Devonian to the Mississippian, the Antler Orogeny affected the western margin of Laurussia, whilst the intra-oceanic Sakmarian-Magnitogorsk arc-trench system, marking the eastern boundary of the Laurussian Plate, was characterised by intermittent orogenic activity through Devonian times. The leading edge of Gondwana collided presumably already during the late Devonian with the Iberian segment of the Appalachian-Variscan arc-trench system. This was followed by the early Carboniferous clockwise rotation of Laurussia and the convergence of the Siberian and Kazakhstan cratons with its eastern margin (Zonenshain et al., 1987b).

The late Palaeozoic collision of Gondwana and Laurussia, and their suturing along the Appalachian-Variscan fold belts spanned some 90 Ma from the time initial contacts were established between these two megacontinents during the late Devonian, until the early Permian consolidation of the Appalachians. The suturing process of Kazakhstan/Siberia and Laurussia lasted some 70 Ma from the Westphalian collision of Kazakhstan with the southern parts of the Moscow Platform, until the early Triassic consolidation of the Uralian Orogen.

In the broadest sense, the Hercynian orogenic cycle could thus be considered as spanning Famennian to early Triassic times. However, keeping in mind that the Appalachian-Variscan megasuture evolved in response to the collision of Gondwana and Laurussia, and that the Ural-Taimyr and Kusbass-Tom-Kolyva megasutures developed in response to the collision of the Siberian and Kazakhstan cratons and their docking to the eastern margin of Laurussia, a distinction should be made between a Hercynian and an Uralian orogenic cycle, despite their broad synchroneity. Although initial contacts were established between Gondwana and Laurussia during the late Devonian and earliest Carboniferous, giving rise to the Bretonian Orogeny, their super-collision commenced only during the late Visean. Therefore the Hercynian Orogeny sensu stricto, marking the docking and accretion of Gondwana to the southern margin of Laurussia, can be defined as spanning late Visean to Autunian times. In this respect the Acado-Ligerian and the Bretonian diastrophic events should be considered as forming part of the Eo-Hercynian orogenic cycle that was governed by the accretion of Gondwana-derived cratonic terranes to the southern margin of Laurussia.

The statement that Gondwana was docked against Laurussia is based on their drift patterns. During the Devonian Laurussia remained more or less stationary, whereas Gondwana converged with it in a dextral oblique, clockwise rotational mode. After initial contacts were established between these two megacontinents during the Famennian in the area of Northwest Africa and Iberia, Laurussia also began to rotate clockwise and drifted 8° northward during the early Carboniferous. This northward movement ceased at the transition from the early to the late Carboniferous. However, clockwise rotation of Laurussia and Gondwana continued during the late Carboniferous to early Permian consolidation of the Hercynian megasuture and amounted to a total of 54°.

Following the early Permian locking of the Hercynian subduction system, Pangea underwent an anti-clockwise rotation of about 20° during the late early and late Permian. This fundamental change in the drift pattern of Gondwana and Laurussia can be attributed to a first order suture progradation from the Hercynian subduction system to the American, Antarctic and Australian margin of Pangea. (Re-?)Activation of this Pacific-type subduction system testifies to continued, if not accelerated, sea-floor spreading within Panthalassa (Proto-Pacific).

The interaction between the eastern margin of Laurussia, Siberia and Kazakhstan was in part controlled by the drift and rotation patterns of Laurussia and in part by the motion of Siberia and Kazakhstan.

The configuration of asthenospheric convection systems, and the dynamics of their changes, governing the observed drift patterns of Gondwana, Laurussia, Kazakhstan and Siberia, and ultimately the fundamental plate boundary re-organization after Pangea was assembled, are a matter of speculation that goes beyond the scope of this paper. Yet, these convection systems appear to provide, possibly in an indirect way, the driving mechanism for plate motions (McKenzie, 1969; Pavoni, 1985, 1988) and for the sum total of lithospheric stresses that governed the evolution and destruction of sedimentary basins and fold belts (Bott and Kuznir, 1984; Dewey, 1988; Park, 1988). The drift of the continents and their interaction is responsible for palaeoclimatologic changes that control to a large extent depositional systems and also faunal and floral provincialities.

With the late Silurian suturing of Laurentia-Greenland and Fennosarmatia the Laurussian megacontinent was formed. During the Devonian and much of the Carboniferous Laurussia played the role of an independent lithospheric plate. However, during the Hercynian and Uralian orogenic cycles, Laurussia gradually lost its identity as a consequence of its integration into the Pangea supercontinent. The post-Hercynian plate boundary re-organization paved the way for the Mesozoic back-up of Pangea. Early phases of this process are already evident during the late Permian in the form of the development of the Tethys and Gondwana rifts and by the activity along the Norwegian-Greenland Sea rift system.

The evolution of major orogenic systems, along which Arctica, Siberia/Kazakhstan and ultimately Gondwana became sutured to Laurussia, was accompanied by the development of large foreland basins. Development of these basins was in part paralleled by major intra-plate compressional deformations, that had a considerable impact on sedimentation patterns on adjacent cratonic platform areas. Apart from these collision-related phenomena, basin evolution in other parts of Laurussia was governed by an-orogenic intra-plate phenomena such as wrench-faulting, rifting and the decay of thermal anomalies. These had repercussions on relative sea-level changes whereby truly eustatic sea-level changes played an over-printing role.

Some of the late Palaeozoic sedimentary basins of Laurussia contain important hydrocarbon provinces (e.g. Appalachian foreland basin, Anadarko, Midland, Paradox, Michigan, Williston basins, Western Canada carbonate shelf, Timan-Pechora area in the northern Ural foreland, Southern Permian Basin of Europe). These basins developed under a variety of geodynamic settings.

In the following some comments are offered on the geodynamics governing the development of sedimentary basins during the late Palaeozoic evolution of Laurussia and its suturing with Gondwana, Kazakhstan and Siberia. Furthermore, the importance of eustatic and tectonically induced relative sea-level changes will be discussed. The final paragraph deals with the effect of plate motions and tectonics on palaeoclimate and biogeography.

9.1. Foreland Basins and Intra-plate Compression

Amongst the foreland basins that developed during the late Palaeozoic evolution of Laurussia a distinction can be made between fore-arc and retro-arc foreland basins. Fore-arc foreland basins are superimposed on passive margin sedimentary prisms. Upon collision of the respective passive margin with an active arc-trench system, fore-arc thrust-loading of the former shelf areas causes its rapid subsidence

(for instance the Ural foredeep). Retro-arc foreland basins can develop on the cratonward side of a Pacific-type active margin (e.g. Silurian-early Carboniferous Appalachian Basin). Upon collision of the respective active margin with an incoming continent, its setting changes to a Himalayan-type collision zone (Uyeda, 1981), whereby back-arc thrusting can cause rapid subsidence of the retro-arc foreland basin. (for instance, the late Carboniferous-early Permian Appalachian basin; Beaumont et al., 1982; Tankard, 1986).

It is likely that the essentially Devonian southern foreland basin of the Inuitian fold belt developed in a retro-arc setting (Trettin and Balkwill, 1979; Frazier and Schwimmer, 1987) in response to the collision of the Arctic Craton with the northern margin of Laurussia. The evolution of this foreland basin was accompanied by the uplift of the Boothia Arch-Cornvallis fold belt and the Inglefield Arch. These features developed in response to the build-up of intra-plate compressional stresses. During the latest Devonian-earliest Carboniferous Ellesmerian orogenic pulse, large parts of this foreland basin became scooped out by thin-skinned thrust sheets, now forming the Parry Island fold belt; this fold belt has a width of some 180 km (Ziegler, 1988).

The foreland basins associated with the Marathon-Ouachita-Appalachian fold belt developed in a retro-arc position. They had already a long pre-collisional history. Particularly the Appalachian Basin received repeatedly orogen-derived clastics during the Silurian, Devonian and early Carboniferous. Its syn-collisional evolution was accompanied by the accumulation of an enormous thickness of Molasse-type clastics. During the terminal phases of the Alleghanian Orogeny this basin became partly scooped out by thin-skinned thrusting. The Appalachian foreland thrustbelt is some 200 km wide. During the evolution of the Appalachian foreland basin the Cincinnati Arch and the Nashville dome became uplifted repeatedly. The model of Quinland and Beaumont (1984) and Beaumont et al. (1987) relates the uplift of these foreland swells (bulges) to visco-clastic deflections of the lithosphere in response to its loading by the advancing thrust sheets. Alternatively, these features may have developed in response to compressional stresses, that were exerted on the foreland during periods of increased orogenic activity. Such an explanation applies particularly for the uplift of the Adirondack Arch that caused the truncation of the northern parts of the Appalachian foreland basin.

The foreland basins of the Ouachita and Marathon thrust belts are subdivided by major wrench induced, transverse striking, upthrusted blocks such as the Wichita-Amarillo uplift. These features form part of the ancestral Rocky Mountains, that developed in response to intraplate compressional stresses during the Alleghanian orogeny at distances up to 1500 km to the north of the contemporaneous thrust front. Uplift of these high relief structures is related to convergent wrench-reactivation of pre-existing fracture systems during the collision of the South American with the North American craton (Kluth, 1986).

Also the late Carboniferous Variscan foreland basin of Europe can be considered as having evolved in a retro-arc position, at least relative to the Proto-Tethys subduction zone. Its precursor, the tensional Devonian-early Carboniferous Rhenohercynian back-arc basin, was probably partly floored by oceanic crust. These oceanic domains became closed during the early phases of the Variscan Orogeny and thrust-loaded subsidence of their northern shelves commenced at the onset of the late Carboniferous. This was accompanied by an earliest Namurian phase of block-faulting, the na-

ture of which is not yet resolved on the basis of the available seismic data. During the terminal phases of the Variscan Orogeny, the southern parts of the Variscan foreland basin became scooped out by thin-skinned thrust-sheets. At the same time intra-plate compressional stresses caused the partial inversion of Carboniferous grabens in the British Isles at distances up to 600 km to the north of the Variscan thrust front. Compressional foreland stresses may, furthermore, be responsible for the tensional reactivation of pre-existing fault systems in the Norwegian-Greenland Sea, causing the onset of rifting in this area.

In contrast, the Uralian foreland basin, developed in a typical fore-arc position. Its precursor is the passive margin of the Moscow Platform, the Timan-Pechora and Eastern Barents shelves that faced the Sakmarian Ocean. During the early Carboniferous the Sakmarian arc became accreted to Kazakhstan, that converged and ultimately collided with the passive margin of the Moscow Platform during the Westphalian, with the Timan-Pechora margin during the Stephanian and with the Barents Shelf during the early Permian. During the evolution of the Uralian foreland basin intraplate compressional stresses caused the inversion of the Devonian-early Carboniferous Pechora-Kolva rifts and the uplift of the Timan Range at distances up to 500 km to the west of the Ural thrustfront.

It is unclear whether major foreland basins had developed parallel to the Hercynian deformation front in Venezuela and in Northwest Africa. During the late Carboniferous and early Permian, the Sahara Platform became transsected by the Reguibat, Anti-Atlas and El-Biot arches and the inverted Ougarta Trough. Development of these features can be related to the build-up of intra-plate compressional stresses during the late phases of the Hercynian orogenic cycle.

Thrust-loading of the foreland by the advancing nappe systems controls the subsidence of foreland basins. The dimensions of such basins are controlled by the magnitude of the tectonic load and by the flexural rigidity of the foreland lithosphere (Beaumont et al., 1982, 1987; Quinlan and Beaumont, 1984; Tankard, 1986). During late stages of orogenic cycles overthickening of the lithosphere can cause thrust propagation into the foreland basin, thereby causing their partial destruction. Post-orogenic erosional unroofing and isostatic uplifting of orogens affects also their foreland basins and leads to their further destruction.

If the orogen and the foreland are mechanically coupled at the respective A-subduction zone, intra-plate compressional stresses can induce foreland deformations and partial destruction of the associated foreland basins (e.g. Ouachita foreland). In this context it is sometimes difficult to discriminate between foreland structures that can be related to intra-plate compression and flexural foreland bulges (Beaumont 1982, 1987; e.g. Appalachian foreland).

Intra-plate compressional stresses appear to build-up, particularly during periods of increased orogenic activity along the margins of the respective continental craton, to the level at which they can cause broad-scale lithospheric deflections (e.g. earliest Devonian uplift of Transcontinental Arch, Permian uplift of Guayana shield) and/or, through re-activation of pre-existing crustal discontinuities, the upthrusting of smaller fault-controlled features (e.g. early Devonian Hudson Bay uplift, Boothia arch, Canada). Such megascopic deformation can be observed as distances up to 1500 km from the associated orogenic thrust front (see also Ziegler, 1988).

9.2. Rift and Wrench Systems

During the late Palaeozoic evolution of Laurussia the development of rift and wrench-induced basins plays an important role.

Intermittent back-arc extension in the domain of the Variscan geosynclinal system is thought to have governed the Devonian-early Carboniferous evolution of the Rhenohercynian and central Armorican-Saxothuringian basins. Back-arc extension may also be responsible for the development of the Donets rift that was coupled with uplift of the Ukrainian-Voronesh thermal dome. Middle Devonian to early Carboniferous rifting on the eastern margin of Laurussia can, at least in part, be related to back-arc extension in the oceanic Sakmarian back-arc basin. The Carboniferous subsidence of the tensional Sverdrup basin may have been governed by back-arc extension related to the decay of the south-plunging Inuitian subduction system. Late Permian and Triassic rifting in the area of the future West Siberian Basin may be related to gradual the decay of the Ural-Novaya Zemlya-Taimyr subduction system. In the Sverdrup and West-Siberian Basins the initial rifting phase was followed by a thermal sag phase. In the Sverdrup Basin this phase became, however, repeatedly interrupted by activity along the trans-Arctic oblique-slip zone (Ziegler, 1988). Permian decay of the north-plunging Variscan subduction zone may have played a role in the evolution of the oceanic Black Sea Basin and during the Stephanian-Auturian phase of wrench-faulting in the Variscan fold belt and its foreland. The main driving force behind the latter is, however, seen in the dextral translation between Gondwana and Laurussia during the Alleghanian Orogeny.

Devonian to early Carboniferous sinistral movements along the Arctic-North Atlantic megashear, transsecting the Arctic-North Atlantic Caledonides and the Acardian fold belt lead to the development of a string of wrench induced basins in which great thicknesses of predominantly continental clastics accumulated.

Although the magnitude and timing of the postulated sinistral translation between Laurentia-Greenland and Fennosarmatia are palaeomagnetically poorly constrained, the existence of such a mega-shear is compatible with the geological record of the Old Red basins that are associated with it. However, the amount of ± 1000 km of mainly intra-Devonian strike-slip motion, as assumed in our model, is possibly excessive in terms of crustal shortening in the hypothetical Lomonosov fold belt, and could be reduced to a value below the palaeomagnetic margin of error.

During the late Carboniferous and Permian the fracture system of the Arctic-North Atlantic mega-shear became tensionally re-activated in the area of the Norwegian-Greenland Sea. Crustal extension in this area was taken up to the north in the evolving dextral trans-Arctic oblique slip zone. Activity along this wrench and pull-apart system played apparently an important role in the post-orogenic collapse of the Inuitian fold belt.

The late Palaeozoic evolution of the Norwegian-Greenland Sea Rift is difficult to explain and hypotheses advanced here are speculative. The inception of this rift system during the early Namurian and its evolution during the Westphalian may be related to the buildup of compressional intra-plate stresses, paralleling the strike of the Devonian-early Carboniferous Arctic-North Atlantic megashear. Stephanian-Autunian southward propagation of the Norwegian-Greenland Sea Rift, during the post-Variscan dextral translation of Africa relative to Europe, may be related to the development of a tensional stress field at the northwestern termination of the Bay of Biscay fracture zone. However, the late Permian, and even more so the Triassic, rapid southward propagation of the Norwegian-Greenland Sea Rift into the North Atlantic domain and into Northwest Europe remains enigmatic, particularly in view

of an extremely low level of volcanic activity. The postulate that far-field tensional stresses, related to orogenic activities along the margins of Laurussia (Sonoma and Ural Orogeny), may be responsible for the Permo-Triassic evolution of the Norwegian Greenland Sea Rift (Bott, 1982; Bott and Kuznir, 1989; Dewey, 1988), requires re-assessment in the face of the above discussed prevalence of intra-plate compression during periods of orogenic activity along plate margins, a phenomenon that is in keeping with the present-day stress patterns of the globe (see Zoback and Zoback, 1980; Klein and Barr, 1986).

Alternatively, it may be speculated that slowly developing upwelling astenospheric convective systems exerted drag forces on the overlying lithosphere and/or caused the development of deviatoric tension within the lithosphere in response to large scale thermal arching, thus governing the late Permian and Mesozoic evolution of the Norwegian Greenland Sea and the Tethys rift systems (Ziegler, 1988; Dewey, 1988).

9.3. Intra-cratonic Thermal-sag Basins

Models on the subsidence of intra-cratonic basins concentrate on the thermal response of the lithosphere to a heating event and the subsequent decay of the thermal anomaly introduced by the former. The classical rifting models distinguish between a rifting stage, during which the lithosphere is thinned by extension and becomes thermally destabilised by an increase of the mantle heat flow, and a post-rifting stage, during which the mantle heat flow decreases to a normal level, causing cooling and contraction of the lithosphere and the subsidence of a so-called thermal sag basin (McKenzie, 1978; Jarvis and McKenzie, 1980; Watts et al., 1982).

The North Sea represents a classical example of a rift basin that is overlain by a thermal sag basin (Barton and Wood, 1984; Ziegler, 1982, 1988; Sclater et al., 1986).

During the rifting stage increased mantle heat flow is either caused by diapiric or convective upwelling of the asthenosphere (mantle plume, active rifting) or by its passive advection in response to thinning of the lithosphere by stretching, induced by far field stresses (passive rifting).

None of the Late Palaeozoic intra-cratonic basins of Laurussia show all the characteristics of a North Sea-type combination of a syn-rift and post-rift thermal sag-basin.

The southern Permian Basin of Europe subsided presumably in response to the thermal relaxation of the lithosphere that became destabilised during the Stephanian-Autunian phase of wrench-faulting and volcanism. However, a typical rifting stage is not evident and a central graben subparalleling the axis of the thermal-sag basin seems to be missing (Ziegler, 1982, 1988).

The Westphalian to Permian regional subsidence of the Sverdrup Basin can be related to the decay of a thermal anomaly that was introduced during its Late Visean-Namurian rifting phase. The configuration of this rift is, however, unknown (Sweeney, 1977; Stephenson et al., 1987).

Mechanisms governing the subsidence of the oval-shaped to subcircular Michigan, Williston, Hudson Bay and Illinois basins have recently been reviewed by Klein and Hsui (1987) and Quinlan (1987). Subsidence of these typically saucer-shaped basins commenced variably during the Late Cambrian to Late Ordovician and persisted into the Mississippian. The Illinois and Michigan basins be-

came incorporated into the flexural Appalachian foreland basin during the early and late Missisippian respectively.

Geophysical data indicate that the crust has not been perceptibly thinned under the axial parts of these basins, the subsidence of which was accompanied by only minor faulting. However, there is no evidence for an initial rifting phase nor for a major thermal event immediately preceding the onset of basin subsidence. In view of this, Quinlan (1987) concludes that the North Sea model does not apply for the evolution of these four North American intra-cratonic basins.

However, the geometry and internal architecture of these basins suggest that they did develop in response to prolonged regional crustal downwarping that could be explained in terms of lithospheric cooling and contraction. On the other hand, as there is no evidence for pre-subsidence hot-spot activity, the evolution of these basins may be related to a decrease of the mantle heat flow above a "cold spot" caused by downwelling flow in the sub-lithospheric mantle (secondary upper mantle convection?). Subsequent recovery of mantle heat flow to "normal" levels would result in the slow emergence and erosion of such basins, unless they become incorporated into an other subsidence regime such as a flexural foreland basin (M. Doyle, personal communication).

The four North American intra-cratonic basins discussed here, subsided during a time when the Laurentian craton underwent only minor latitudinal drift. This suggests that drag-forces exerted by the convecting asthenosphere on the Laurentian craton remained in balance. In this context it is speculated that during the evolution of these basins Laurentia was the site of converging, downwelling asthenospheric currents, corresponding to cold spots. During the Mississippian, and particularly during the

Pennsylvanian and Permian the Laurentian craton drifted northward and these basins became presumably decoupled from their cold spots, ceased to subside and became gradually uplifted as mantle heat flow regained normal levels. Fission track data from the Williston and Michigan basin indicate that they became uplifted and eroded during the latest Palaeozoic and Mesozoic (Crowley et al., 1985; Crowley and Kuhlman, 1988). The same probably applies also for the Hudson Bay Basins. The Illinoi Basin, on the other hand, became incorporated into the flexural Appalachian foreland basin and there is evidence for late Pennsylvanian to Permian magmatic activity (Klein and Hsui, 1987).

This novel "cold spot" model for thermal sag basins, that does not conform to the classical North Sea model, requires further testing.

9.4. Eustatic and Relative Sea-level Changes

The configuration and dimensions of the Old Red continent, occupying the central parts of Laurussia, was influenced by intra-plate deformations and also by eustatic sea-levels changes (Vail et al., 1977; House, 1983; Johnson et al., 1985). The cyclical Devonian rise in sea level, that extended into the early Carboniferous, is probably of a tectono-eustatic nature. A glacio-eustatic contribution to these sea-level fluctuations cannot be excluded, particularly for the Famennian and early Carboniferous; by this time substantial continental ice sheets had developed in Gondwana. There are, however, important differences in the timing and degree of flooding and emersion of the Fennosarmatian, North American and Sahara platforms, for instance during the early Devonian and the Mississippian. Differences observed in relative sea-level fluctuations in often widely separated basins are probably the result of epirogenetic movements that were induced by tensional and

compressional intra-plate stresses and in places by the introduction of positive and negative thermal anomalies (Johnson et al., 1985, see also Cloetingh et al., 1985).

During the late Carboniferous and Permian, glaciation of southern Gondwana and the possible development of an ice-cap on northern Siberia during the late Permian, are held responsible for major eustatic sea-level changes. At the same time tectono-eustatic sea-level changes can be envisaged in terms as defined by Heller and Angevine (1985). The generally accepted very low stand in global sea-level at the end of the Permian (Schopf, 1974; Forney, 1975; Ross and Ross, 1987) corresponds to a time when the globe was dominated by one hugh ocean basin, Panthalassa (Proto-Pacific). Large parts of this basin were occupied by old oceanic lithosphere that had subsided to its maximum depth. Thus, the global ocean basins had reached a maximum volume. Moreover, the proportion of the surface area of the globe, that was occupied by continental crust, had been reduced by the Hercynian and Uralian continent-to-continent collisions.

Despite these important glacio-and tectono-eustatic sea-level changes, which are difficult to quantify, major late Carboniferous and Permian transgressions and regressions, causing important changes in the outlines of intra-cratonic basins and the shape of the Central Laurussian landmass, were apparently controlled to a large extent by tectonic phenomena.

An example in point is the evolution of the Arctic-Barents Shelf, which is related to the collapse of the early Carboniferous Inuitian-Lomonosov fold belt, the subsidence of the post-orogenic Sverdrup back-arc basin and activity along the Norwegian-Greenland Sea- and Trans Arctic rift-wrench system. Furthermore, rift-induced subsidence of the Norwegian-Greenland Sea area, combined with thermal subsidence of the Northern and Southern Permian Basins of Europe, caused the late Permian southward transgression of the Arctic Seas. On the other hand, post-orogenic uplift of the Appalachian Orogen and its forelands underlies the post-Autunian regional regression in the domain of the Central Atlantic and the Gulf of Mexico. Broad-scale, probably collision-related lithospheric deflections may be held responsible for the Chesterian uplift of the Transcontinental Arch and the late Permian uplift of the Lomonosov High.

In short, intra-plate tectonic activity in the cratonic parts of Laurussia controlled to a large extent the pattern of transgressions and regressions, whereby glacio-eustatic and tectono-eustatic sea-level changes played an overprinting role. It is suspected that the vertical amplitude of intra-plate deformations considerably exceeds the amplitude of eustatic sea-level fluctuations.

Finally, changes in the geoïd-shape, induced by changes in the deep mantle convection systems, as a consequence of the insulating effect of the Pangea super-continent, may have contributed to the late Permian and early Triassic regression that is evident in North America, northern Africa and Europe (Pavoni, 1985, 1988).

9.5. Palaeoclimate and Palaeobiogeography

The Old Red landmass occupied throughout Devonian times an equatorial position. Its dimensions were largest during the Gedinnian-Siegenian and became gradually reduced due to the flooding of large parts of the North American and Fennosarmatian platforms. However, with the gradual evolution of the Inuitian and Appalachian orogens, and the outbuilding of major deltas into their foredeeps, the land area of the Old Red Continent increased again.

The sedimentary record of this landmass indicates that it was characterised by a hot, at least seasonally arid climate (Woodrow et al., 1973; Johnson, 1981). Rainfall pattern were probably heavily influenced by the rising fold belts of the Inuitian, Appalachian-Variscan and Antler orogens. These topographic features acted presumably as seasonal precipitation barriers that thus enhanced the aridity of the Old Red Continent. This applies particularly to its western shelf seas, that contain major evaporitic deposits. The huge deltaic complexes, that prograded during the middle and late Devonian into the southern foreland basin of the Inuitian fold belt, must have been associated with major stream systems. Similar, streams originating on the lee-ward side of the ancestral Appalachians, supplied the middle and late Devonian Catskill deltas.

In the Devonian to early Carboniferous wrench-induced intramontane basins, that subsided in the area of the Arctic North Atlantic Caledonides, the occurrence of thick conglomeratic series and cyclical lacustine deposits, containing fresh water fish and Tetrapods, testifies for high erosion rates, clastic transport by torrential and braided streams and sufficient rainfall to maintain perennial lakes (Hamilton and Trewin, 1988).

The Old Red Continent was one of the important landmasses on which terrestial plant and animal life evolved during the Devonian (Johnson, 1981). Faunal and floral provincialism was evident during the early and middle Devonian. With the Famennian establishment of initial contacts between Gondwana and Laurussia terrestrial biotas became cosmopolitan (Young, 1987).

Late Carboniferous and early Permian suturing

of Gondwana and Laurussia, and the Permian accretion of Siberia/Kazakhstan to the newly formed Pangea supercontinent, as well as the late Permian drift of its northern parts into Arctic waters, had severe repercussions on the faunal provinciality of shallow marine warm water faunas. Early Carboniferous cosmopolitan faunal patterns were disrupted by the Namurian closure of the circum-equatorial seaway through North Africa. Coldwater barriers, both in the Arctic and in Antarctica, impeded the dispersal of tropical shelf faunas. This caused the development of new warm water faunal provinces (Ross, 1973; Ross and Ross, 1985b; Gobbet, 1973).

At the same time the development of a widening land bridge between Gondwana and Laurus-

sia permitted a broad exchange of terrestial faunas and floras, whereby the late Permian increasing aridity of equatorial zones may have provided certain constraints (Parrish et al., 1986; Rowley et al., 1985; Romer, 1975; Allen and Dineley, 1988).

The late Permian presence of continental ice caps in Arctic and Antarctic regions caused presumably a concentration of the world's air pressure belts towards the equator, thus creating a shorter distance between the zones of high and low pressure than is now the case. Resulting wind systems had presumably higher average velocities than now (Glennie, 1986). This probably contributed to the late Permian aridity of the central parts of Pangea.

Postscript

The task of unscrambling the late Palaeozoic plate motions, that underlay the step-wise assembly of Laurussia and ultimately its integration into Pangea, demands a multidisciplinary approach. No single geological or geophysical discipline is able to provide a data base that, on its own, is able to answer all the questions that are raised by the data bases of the other earth science disciplines. Thus, forces must be joined in an effort to find a common denomination that honours as far as possible the data at hand.

A synthesis, as the one presented in this volume, is obviously based on arbitrary decisions of the author. In the face of conflicting data, he has to decide which evidence must be given priority. In doing so, he should be guided by geodynamic principles, as far as they are known and understood at present. For instance, development of a major clastic filled foreland basin is not compatible with the concept of rifting and the separation of two continents that have collided shortly before (e.g. early and middle Devonian Appalachian domain). Furthermore, it is unlikely that continents drifting past each other and ultimately converging with each other on a collisional course, show drift patterns involving erratic zig-zag tracks. To this end, drift patterns of major cratons and continental fragments, as presented in this study, have been evaluated by a computer animation.

Yet, the author painfully realises that the data bases available to him are still incomplete and that many questions had to remain unanswered. Thus the model present in this volume for the evolution of Laurussia is not unique and has to remain a hypothesis that will require modification as new data become available.

It is hoped that publication of this study will stimulate others to pursue this approach to the understanding of the evolution of the lithosphere. Only by pooling knowledge and experience will we be able to advance the Earth Sciences. Specialists active in a given geoscience discipline must communicate with specialists of other disciplines.

The ultimate goal is the understanding of the past, present and future behaviour of our beautiful, life supporting Blue Planet.

P.A. Ziegler
Binningen, Switzerland
January 1989

References

Adamia, Sh.A., Chkhotua, T., Kelelia, M., Lord Kipanidze, M. and Shavishvili, I., 1981 – Tectonics of the Caucasus and adjoining regions: implications for the evolution of the Tethys Ocean. J. Struct. Geol., 3 (4): 437–447

Aifa, T., 1987 – Paléomagnetisme en zones de collision: Déformations recentes dans l'arc Thyrrhenian et racourcissement crustal hercynien en Afrique. Mém. Sc. Terre Univ. Curie, Paris, 87–29: 183 p

Aleinikov, A.L., Bellavin, O.V., Bulashevich, Yu.P., Tarvin, I.F., Maksimov, E.M., Rudkevich, M.Ya., Nalivkin, V.D., Shablinskaya, N.V. and Sutkov V.S., 1980 – Dynamics of the Russian and West Siberian Platforms. In: A.W. Bally, P.L. Bender, T.R. McGretchin and R.I. Walcott (eds.). Dynamics of plate interiors. Geodynamics Series, vol. 1. Am. Geophys. Union; Geol. Soc. Am.: 53–72

Allen, J.R.L., 1985 – Marine to fresh water: the sedimentology of the interrupted environmental transition (Ludlow-Siegenian) in the Anglo-Welsh Basin. In: W.G. Chaloner and J.D. Lawson (eds.), Evolution and environment in the Late Silurian and Early Devonian, Phil. Trans. R. Soc. Lond., B309: 85–104.

Allen, J.R.L., Dineley, D.L. and Friend, P.F., 1967 – Old Red sandstone basins of North America and Northwest Europe. In: D.H. Oswald (ed.), International Symposium on the Devonian System, Alberta Soc. Petrol. Geol., vol. 1: 69–98

Allen, K.C. and Dineley, D.L., 1988 – Mid-Devonian to mid-Permian floral and faunal regions and provinces. In: A.L. Harris and D.J. Fettes (eds.). The Caledonian-Appalachian orogen. Geol. Soc. Spec. Publ. 38: 531–548

Amsden, T.W., Caplan, W.M., McGlassson, E.H., Hilpman, P.L., Rowland, T.L. and Wise, O.A. jr., 1967 – Devonian of the southern mid continent area, United States. In: D.H. Oswald (ed.), International Symposium on the Devonian System, Calgary, Alberta Soc. Petrol. Geol., vol. 1: 913–933

Anderton, R., Bridges, P.H., Leeder, M.R. and Sellwood, B.W., 1979 – A dynamic stratigraphy of the British Isles: a study in crustal evolution. George Allen & Unwin, London, 301 p

Argyriades, I., de Graciansky, P.C., Marcoux, J. and Ricou, L.E., 1980 – The opening of the Mesozoic Tethys between Eurasia and Arabia-Africa. In: Géologie des chaines alpines issues de la Tethys. Mem. BRGM 115: 198–214

Armin, R.A., 1987 – Sedimentology and tectonic significance of Wolfcampian (Low. Perm.) conglomerates in the Pedregosa basin: southeastern Arizona, southwestern New Mexico and northern Mexico. Geol. Soc. Am. Bull. V. 99: 42–65

Arthaud, F. and Matte, P., 1977 – Late Palaeozoic strike-slip faulting in southern Europe and North Africa: results of a right-lateral shear zone between the Appalachians and the Urals. Geol. Soc. Am. Bull. 88: 1305–1320

Arthyushkov, E.V. and Baer, M.A., 1983 –

Mechanism of continental crust subsidence in fold belts, the Urals, Appalachians and Scandinavian Caledonides. Tectonophysics, 100: 5–42

Autran, A. and Cogné, J., 1980 – La zone interne de l'orogène Varisque dans l'ouest de la France et sa place dans le developpement de la chaîne Hercynienne. Mem. BRGM, 108: 90–111

Autran, A. and Dercourt, J. (eds.), 1980 – Evolution géolgique de La France. Mem. BRGM, 107: 335

Badham, J.P.N., 1982 – Strike-slip orogens – An explanation for the Hercynides. J. Geol. Soc. Lond, 139: 493–504

Baker, A.J. and Gayer, R.A., 1985 – Caledonide-Appalachian tectonic analysis and evolution of related Oceans. In: R.A. Gayer (ed.), The tectonic evolution of the Caledonide-Appalachian Orogen, Fried. Vieweg & Son, Braunschweig/Wiesbaden, 126–165

Balkwill, H.R. and Fox, F.G., 1982 – Incipient rift zone, Western Sverdrup Basin, Arctic Canada. In: A.F. Embry and H.R. Balkwill (eds.), Arctic Geology and Geophysics, Can. Soc. Petrol. Geol. Mem. 8: 171–186

Barton, P. and Wood, R., 1984 – Tectonic evolution of the North Sea basin: crustal stretching and subsidence. Geophys. J. Roy. Astr. Soc. 79: 987–1022

Bassett, H.G. and Stout, J.G., 1967 – Western Canada. In: D.H. Oswald (ed.), International Symposium of the Devonian System, Alberta Soc. Petrol. Geol., vol. 1: 717–752

Beauchamp, J. and Izart, A., 1987 – Early Carboniferous basins of the Atlas-Meseta domain (Morocco): Sedimentary model and geodynamic evolution. Geology, 15: 797–800

Beaumont, C., Keen, C.E. and Boutillier, R., 1982 – A comparison of foreland and rifted margin sedimentary basins. Phil. Trans. Roy. Soc. London, 305A: 295–317

Baumont, C., Quinlan, G.M. and Hamilton, J.,

1987 – The Alleghanian orogeny and its relationship to the evolution of the eastern interior, North America. In: C. Beaumont and A.J. Tankard (eds.). Sedimentary basins and basin-forming mechanisms. Can. Soc. Petrol. Geol. Mem. 12: 425–447

Beckinsale, R.D., Reading, H.G. and Rex, D.C., 1975 – Potassium-Argon ages from basic dykes from Eastern Finnmark: stratigraphic and structural implications. Scott J. Geol. 12: 51–65

Behr, H.J., Engel, W. and Franke, W., 1982 – Variscan wildflysch and nappe tectonics in the Saxothuringian zone (northeast Bavaria, West Germany). Am. J. Sci. 282: 1438–1470

Behr, H.J., Engel, W., Franke, W., Giese, P. and Weber, K., 1984 – The Variscan belt in central Europe: main structure and geodynamic implications: open questions. Tectonophysics, 109: 15–40

Bell, J.S., 1973 – Late Paleozoic orogeny in the Northern Yukon. In: J.D. Aitken and D.J. Glass (eds.). Proceedings of the Symposium on the Geology of the Canadian Arctic, Saskatoon, May 1973, Geol. Ass. Can. and Can. Soc. Petrol. Geol.: 25–38

Bernard-Griffiths, J., Cantagrel, J.M. and Duthon, J.C., 1977 – Radiometric evidence for an Acadian tectonometamorphic event in the Western Massif Central Francais. Contr. Miner. Petrol., 61: 199–212

Bird, K.J. and Molenaar, C.M., 1986 – The framework geology of the North Slope of Alaska as related to oil/source rock correlations. In: L. Magoon and G. Claypool (eds.). Alaska North Slope oil/source rock correlation study, Am. Assoc. Petrol. Geol., Studies in Geology Series No. 20.

Bird, K.J., Connor, C.L., Tailleur, I.L., Silberman, M.L. and Christie, J.L., 1978 – Granite on the Barrow Arch, Northeast NPR-A. In: K.M. Johnson (ed.). The U.S. Geological Survey in Alaska: Accomplishments during 1977, US. Geol. Surv. Circ. 772 B, B24-B25

Birkenmajer, K., 1981 – The Geology of Svalbard, the western part of the Barents Sea, and the continental margin of Scandinavia. In: A.E.M. Nairn, M. Churkin jr. and F.G. Stehli (eds.). The Ocean Basins and Margins, vol 5, The Arctic Ocean, Plenum Press, New York, 265–329

Bluck B.J., 1978 – Sedimentation in a late orogenic basin: the Old Red Sandstone of the Midland Valley. In: D.R. Bowes and B.E. Leake (eds.). Crustal evolution in NW Britain and adjacent regions, Geol. J. Spec. Issue, 10: 249–278

Bluck, B.J., 1983 – Role of the Midland Valley of Scotland in the Caledonian Orogeny. Trans. Roy. Soc. Edinburgh: Earth Sci. 74: 119–136

Bluck, B.J., 1984 – Pre-Carboniferous history of the Midland Valley of Scotland. Trans. Roy. Soc. Edinburgh: Earth Sci., 75: 275–295

Blythe, A.E., Sugar, A. and Phipps, S.P., 1988 – Structural profiles of the Ouachita Mountains western Arkansas. Am. Assoc. Petrol. Geol. Bull. 72(7): 810–819

Bott, M.H.P., 1982 – Origin of the lithospheric tension causing basin formation. In: Sir Peter Kent, M.H.P. Bott, D.P. McKenzie and C.A. Williams (eds.). The evolution of sedimentary basins. Phil. Trans. R.Soc. Lond., A305: 309–324

Bott, M.H.P. and Kuznir, N.J., 1984 – The origin of tectonic stress in the lithosphere. Tectonophysics, 105: 1–13

Boucot, A.J., Dewey, J.F., Dineley, D.L., Fletcher, R., Fyson, W.K., Griffin, J.G., Hickox, C.F., McKerrow, W.S. and Ziegler, A.M., 1974 – Geology of the Arisaig area, Antigonish County, Nova Scotia. Geol. Soc. Am. Spec. Paper 139, 191 p

Bouillin, J.P. and Perret, M.F., 1982 – Datation par Conodonts du Carbonifère inférieur et mise en évidence d'une phase tectonique bretonne en Petit Kabylie (Algérie). C.R. Acad. Sc. Paris, Ser. II, 295: 47–50

Bourque, P.A., Amyot, G., Desrochers, A., Gigane, H., Gosselin, C., Lachambre, G. and Laliberte, J.Y., 1986 – Silurian and Lower Devonian Reef and Carbonate complexes of the Gaspé Basin, Quebec. A summary. Can. J. Earth Sci., 34 (4): 452–489

Bradley, D.C., 1982 – Subsidence in Late Paleozoic basins in the Northern Appalachians. Tectonics, 1 (1): 107–123

Bradley, D.C., 1983 – Tectonics of the Acadian orogeny in New England and adjacent Canada. J. Geol., 91: 381–400

Braun, W.K. and Mathison, J.E., 1986 – Mid-Givetian events in Western Canada: The Dawson Bay – Watt Mountain – "Slave Point" Interlude. Bull. Can. Petrol. Geol., 34 (4): 426–451

Bukovics, C. and Ziegler, P.A., 1985 – Tectonic development of the mid-Norway continental margin. Marine Petrol. Geol., 2 (1): 2–22

Bullard, E.C., Everett, J.E. and Smith, A.G., 1965 – The fit of the Continents around the Atlantic. In: A symposium on Continental Drift, Phil. Trans. Roy. Soc. London, Ser. A 258: 41–51

Burchfiel, B.C. and Davis, G.A., 1972 – Structural framework and evolution of the southern part of the Cordilleran orogen, western United States. Am. J. Sci., 272: 97–118

Burchfiel, B.C. and Davis, G.A., 1975 – Nature and controls of Cordilleran orogenesis, western United States: extension of an earlier synthesis. Am. J. Sci. 275: 363–396

Burg, L.P., Leyreloup, A., Marchand, J. and Malle, Ph., 1984 – Inverted metamorphic zonation and large-scale thrusting in the Variscan belt: an example in the French Massif Central. In: D.H.W. Hutton and D.J. Sanderson (eds.). Variscan tectonics of the North Atlantic Region. Geol. Soc. Lond., Spec. Publ. No. 14: 47–61

Burrowes, O.G. and Krause, F.F., 1987 – Overview of the Devonian System: subsurface of Western Canada Basin. In: F.F. Krause and

O.G. Burrowes (eds.). Devonian Lithofacies and Reservoir Styles in Alberta, 2nd Internat. Symp. on the Devonian System, Calgary, Alberta, Canada, August 17–20th, 1987, 1–20

Caputo, M.V., 1985 – Late Devonian glaciation in South America. Palaeogeogr., Palaeoclim., Palaeoecol., 51: 291–317

Caputo, M.V. and Crowell, J.C., 1985 – Migration of glacial centers across Gondwana during Paleozoic era. Geol. Soc. Am. Bull. 96: 1020–1036

Carlier, G., Gardin, G., Laubacker, G., Marocco, R. and Mégard, F., 1982 – Present knowledge of the magmatic evolution of the eastern Cordillera of Peru. Earth Sci. Rev. 18: 253–283

Carls, P., 1983 – La zona Asturoccidental-Leonesa en Aragon y el Macizo del Ebro como prolongacion del Macizo Cantabrico. In: Libro Jubilar J.M. Ramos, Tomo III, contribuciones sobre temas generales, 11–32

Carls, P. and Lages, H., 1983 – Givetium und Ober-Devon in den östlichen Iberischen Ketten (Spanien). Z. dt. geol. Ges., 134: 119–142

Carls, P., 1988 – The Devonian of Celtiberia (Spain) and Devonian Palaeogeography of SW Europe. In: N.J. McMillan, A.F. Embry and D.J. Glass (eds.). Devonian of the World. Can. Soc. Petrol. Geol. Mem. 14 , Vol.1: 421–466

Carter, D.G. and Pickerill, R.K., 1985 – Lithostratigraphy of the Late Devonian-Early Carboniferous Horton Group of the Moncton subbasin, southern New Brunswick. Maritime Sediments and Atlantic Geology, 21(1–3): 11–24

Castellarin, A. and Vai, G.B., 1981 – Importance of Hercynian tectonics within the framework of the southern Alps. Jl. Struct. Geol. 3 (4): 477–486

Cawood, P.A., Williams, H., O'Brien, S.J. and O'Neill, P.P., 1988 – Field trip guide book, trip A–1, geological cross-section of the Appalachian orogeny; St. John's 1988. Geol. Assoc. Canada, 160 p

Cazes, M., Torreilles, G., Bois, C., Damotte, B., Galdeano, A., Hirn, A., Mascle, A., Matte, Ph., Pham van Ngoc and Raoult, J.F., 1985 – Structure de la croûte hercynienne du Nord de la France: premiers rérsultats du profil ECORS. Bull. Soc. géol. France, 1985 (8) I (6): 925–941

Chablinskaia, N.V., Koulikov, P.K., Nalivkin, V.D., Pogrebitsky, Yu.E., Polkin, Ia.I., Rostovtsev, N.N., Rudkevich, M.Ia. and Smirnov, V.G., 1982 – La platforme de la Siberie occidentale. In: Tectonics of Europe and adjacent areas, Variscides, Epi-Paleozoic platformes, Alpides. Explanatory note to the International tectonic map of Europe and adjacent areas. Scale 1:2,500,000, "Nauka" Publ. House, Moscow, 265–276

Chacón, J., Oliveira, V., Ribeiro, A. and Oliveira, J.T., 1983 – La Estrutura de la zona de Ossa Morena. In: Libro Jubilar J.M. Ramos, Geologia de Espana, Tomo I, Inst. Geol. Minero Espana Publ., 490–504

Chorlton, C.B. and Dallmeyer, R.D., 1986 – Geochronology of Early to Middle Paleozoic tectonic development in the southwest Newfoundland Gander zone. J. Geol., 94: 67–89

Churkin, M., jr. and Trexler, J.H., jr., 1981 – Continental plates and accreted oceanic terranes in the Arctic. In: A.E.M. Nairn, M. Churkin jr. and E.G. Stehli (eds.). The ocean basins and margins, vol. 5. The Arctic Ocean, Plenum Press (New York and London), 1–20

Churkin, M., jr., Soleimani, G., Carter, G. and Robinson, R., 1981 – Geology of the Soviet Arctic: Kola Peninsula to Lena River. In: A.E.M. Nairn, M. Churkin, jr. and F.G. Stehlin (eds.). The ocean basins and margins. Vol. 5. The Arctic Ocean, Plenum Press (New York and London), 331–375

Clement, J.H., 1986 – Cedar Creek: A significant paleotectonic feature of the Williston Basin. In: J.A. Peterson (ed.) Paleotectonics and sedimentation in the Rocky Mountain Region, United States, Am. Assoc. Petrol. Geol. Mem. 41: 213–240

Cloetingh, S., McQueen, H. and Lambeck, K., 1985 – On a tectonic mechanism for regional sea level variations. Earth Planet. Sci. Lett., 75: 157–166

Cogné, J. and Lefort, J.P., 1985 – The Ligerian Orogeny: a Proto-Variscan event, related to the Siluro-Devonian evolution of the Tethys I ocean. In: D.G. Gee and B.A. Sturt (eds.). The Caledonide Orogen-Scandinavia and Related Areas (part 2). John Wiley & Sons, 1185–1195

Collinson,C., 1967 – Devonian of the North Central Region, United States. In: D.H. Oswald (ed.). International Symposium on the Devonian System Calgary, Alberta Soc. Petrol. Geol. Vol. I: 933–973

Conrad, J., Massa, D. and Weyant, M., 1987 – Late Devonian regression and Early Carboniferous transgression on the North African platform. Ann. Soc. géol. Belgique, 109: 113–122

Cook, F.A., 1983 – Some consequences of palinspastic reconstruction in the southern Appalachians. Geology, 11: 89–98

Cook, F.A., Albaugh, D.S., Brown, L.D., Kaufman, S., Oliver, J.E. and Hatcher, R.D., jr., 1979 – Thin-skinned tectonics in the crystalline southern Appalachians: COCORP seismic-reflection profiling of the Blue Ridge and Piedmont. Geology 7: 563–567

Cook, F.A., Brown, L.D., Kaufman, S., Oliver, J.E. and Petersen, T.A., 1981 – COCORP seismic profiling of the Appalachian orogen beneath the coastal plain of Georgia, Part I. Geol. Soc. Am. Bull. 92: 738–748

Cook, T.D. and Bally, A.W., 1975 – Stratigraphic Atlas, North and Central America. Shell Oil Co., Exploration Department, Houston, Texas

Coward, M.P. and Enfield, M.A., 1987 – The structure of the West Orkney and adjacent basins. In: J. Brooks and K. Glennie (eds.). Petroleum geology of North West Europe. Graham and Trotman, vol. 2, 687–696

Cox, L.R.M. and Fortey, R.A., 1988 – Lower Paleozoic facies and faunas around Gondwana. In: M.G. Audley-Charles and A. Hallam (eds.). Gondwana and Tethys. Geol. Soc. Lond., Spec. Publ. 37; 183–200

Craig, L.C. and Waite-Connor, C., 1979 – Paleotectonic investigations of the Mississippian system in the United States, Part 1: introduction and regional analysis of the Mississippian system. Geol. Survey, Professional Paper 1010, Washington, 369 p

Crasquin, S., Lethiers, F. and Mansy, J.-L., 1986 – Modification du cortège ostracodique à la limite Dévono-Carbonifère dans l'Ouest canadien: une conséquence de l'orogenèse Antler? Bull. Soc. géol. France, (8) II 5: 735–740

Crowley, K.D. and Kuhlman, S.L., 1988 – Apatite thermochronometry of Western Canadian Shield: implications for the origin of the Williston Basin. Geophys. Research Lett. 15: 221–224

Crowley, K.D., Ahern, J.L. and Naeser, C.W., 1985 – Origin and epeirogenic history of the Williston Basin: evidence from fission-track analysis of apatite. Geology 13: 620–623

Cutt, B.J. and Laving, J.G., 1977 – Tectonic elements and geologic history of the South Labrador and Newfoundland Continental Shelf, Eastern Canada. Can. Soc. Petrol. Geol. Bull. 25: 1037–1058

Dallmeyer, R.D., 1978 – 40Ar/39Ar incremental-release ages of hornblende and biotite across the Georgia Inner Piedmont, their bearing on late Paleozoic-early Mesozoic tectonothermal history. Am. J. Science, 278: 124–149

Dallmeyer, R.D., 1982 – 40Ar/39Ar ages from the Narragansett basin and southern Rhode Island basement terrane: their bearing on the extent and timing of Alleghenian tectonothermal events in New England. Geol. Soc. Am.

Bull., 93: 1118–1130

Dallmeyer, R.D., 1987 – 40Ar/39Ar age of detrital muscovite within Lower Ordovician sandstone in the coastal plan basement of Florida: implications for west African terrane linkages. Geology, vol. 15: 998–1001

Dallmeyer, R.D., Wright, J.E., Secor, D.T. and Snoke, A.W., 1986 – Character of the Alleghanian orogeny in the southern Appalachians, Part II. Geochronological constraints on the tectonothermal evolution of the eastern Piedmont in South Carolina. Geol. Soc. Am. Bull. 97: 1329–1344

Dallmeyer, R.D. and Keppie, J.D., 1987 – Polyphase Late Paleozoic tectonothermal evolution of the southwestern Meguma terrane, Nova Scotia: evidence from 40Ar/39Ar mineral ages. Can. J. Earth Sci., 24: 1242–1252

Dallmeyer, R.D. and Villeneuve, M., 1987 – 40Ar/39Ar mineral age record of polyphase tectonothermal evolution in the southern Mauretanide orogen, southern Senegal. Geol. Soc. Am. Bull., 98: 602–611

Daly, L. and Irving, E., 1983 – Paléomagnetism des roches carbonifères du Sahara Central; analyse et aimanations juxtaposées; configuration de la Pangée. Ann. Geophys. 1 (3): 207–216

Dankers, P., 1982 – Implications of Early Devonian poles from the Canadian Arctic Archipelago for the North American apparent polar wander path. Can. J. Earth Sci., 19: 1802–1809

Dewey, J.F., 1988 – Lithospheric stress, deformation and tectonic cycles: the disruption of Pangea and the closure of Tethys. In: M.G. Audley-Charles and A. Hallam (eds.). Gondwana and Tethys. Geol. Soc. Lond., Spec. Publ. 37: 23–40

Dimian, M.V., Gray, R., Stout, J. and Wood, B., 1983 – Hudson Bay Basin.In: A.W. Bally (ed.). Seismic expression of structural styles a picture and work atlas, Am. Assoc. Petrol. Geol.-Studies in Geology, Series No.

15, vol. 2, 2.2.4.1–4

Donaldson, A.C. and Shumaker, R.C., 1981 – Late Paleozoic Molasse of Central Appalachians. In: A.D. Miall (ed.), Sedimentation and Tectonics in Alluvial Basins, Geol. Ass. Canada, Special Paper 23: 99–124

Dorobek, S.L. and Read, J.F., 1986 – Sedimentology and basin evolution of the Siluro-Devonian Helderberg group, Central Appalachians. J. Sed. Petrology, 56 (5): 601–613

Dubé, T.E., 1988 – Tectonic significance of Upper Devonian igneous rocks and bedded Barite, Roberts Mountain Allochthon, Nevada, USA. In: N.J. McMillan, A.F. Embry and D.J. Glass (eds.). Devonian of the World. Can. Soc. Petrol. Geol. Mem. 14, Vol. 2: 235–251

Dutro, J.T. jr., 1981 – Geology of Alaska bordering the Arctic Ocean. In: A.E.M. Nairn, M. Churkin jr. and F.G. Stehli (eds.). The Ocean Basins and Margins, vol. 5, The Arctic Ocean, Plenum Press (New York, London), 21–36

Dymkin, A.M.D., Ivanov, S.N., Kamaletdinov, M.A., Popov, B.A., Puchkow, V.N., Zoloev, K.K. and Fishman, M.V.F., 1984 – Geology of the Urals. In: Geology of the USSR, Proc. 27th Int. Geol. Congr. Col. 01, vol. 1: 61–71

Eardley, A.J., 1951 – Structural Geology of North America. Harper & Brothers Publ. New York, 624 p

Eichmüller, K. and Seibert, P., 1984 – Faziesentwicklung zwischen Tournai und Westphal D. im Kantabrischem Gebirge. Z.dt. geol. Ges. 135: 163–191

Eisbacher G.H., 1983 – Devonian Mississippian sinistral transcurrent faulting along the cratonic margin of Western North America: a hypothesis. Geology, 11: 7–10

Embry, A.F. and Klovan, J.E., 1976 – The Devonian clastic wedge of the Canadian Arctic Archipelago. Bull. Can. Soc. Petrol. Geol., 24: 489–622

Enfield, M.A. and Coward, M.P., 1987 – The

structure of the West Orkney Basin, northern Scotland. J. Geol. Soc. Lond., 144 (6): 871–884

Engel,W., Franke, W. and Langenstrassen, F., 1983 – Palaeozoic sedimentation in the Northern branch of the Mid-European Variscides – Essay of an interpretation. In: H. Martin and F.W. Eder (eds.). Intracontinental fold belts, Springer Verlag (Berlin, Heidelberg), 9–42

Ettensohn, F.R., 1985 – The Catskill Delta complex and the Acadian Orogeny: a model. In: D.L. Woodrow and W.D. Sevon (eds.). The Catskill Delta Geol. Soc. Am. Spec. Paper, 201: 39–49

Ettensohn, F.R., 1987 – Rates of relative motion during the Acadian Orogeny based on the spatial distribution of Black Shales. J. Geol., 95: 572–582

Floyd, P.A., and Leveridge, B.E., 1987 – Tectonic environment of the Devonian Gramscatho basin, South Cornwall: framework mode and geochemical evidence from turbidite sandstones. J. Geol. Soc. Lond., 144: 531–542

Forney, G.G., 1975 – Permo-Triassic sea level change. Jl. Geol. 83: 773–339

Fortey, R.A., 1984 – Global earlier Ordovician transgressions and regressions and their biogeological implications. In: D.L. Burton (ed.). Aspects of the Ordovician system. Palaeontological contributions from the University of Oslo No. 295. Universitetsforlaget, Oslo-Bergen-Stavanger-Tromsø, 37–50

Fox, F.G., 1985 – Structural geology of the Parry Islands fold belt. Bull. Can. Soc. Petrol. Geol., 33 (3): 306–340

Fralick, P.W. and Schenk, P.E., 1981 – Molasse Deposition and Basin Evolution in a wrench tectonic setting; the late Paleozoic, East Cumberland Basin, Maritime Canada. In: A.D. Miall (ed.). Sedimentation and Tectonics in Alluvial Basins, Geol. Ass. Can. Special Paper 23: 77–97

Francis, E.H., 1978 – Igneous activity in a fractured craton: Carboniferous volcanism in Northern Britain. In: D.R. Bowes and B.E. Leake (eds.) Crustal evolution in N.W. Britain and adjacent regions, Geol. J. Lond., Spec. Issue, 10: 279–296

Francis, E.H., 1988 – Mid-Devonian to Early Permian volcanism: Old World In: A.L. Harris and D.J. Fettes (eds.). The Caledonian-Appalachian Orogen Blackwell Sci. Publ. Geol. Soc. Lond., Spec. Publ. 38: 573–584

Franke, W., 1988 – Tectonostratigraphic units in the Variscan belt of Central Europe. In: R.D. Dallmeyer (ed.). Terranes in the Circum-Atlantic Paleozoic orogens. Geol. Soc. Am. Spec. Paper 230 (in press)

Franke, W. and Engel, W., 1986 – Synorogenic sedimentation in the Variscan Belt of Europe. Bull. Soc. géol. France, (8) II (1): 25–33

Franke, W., Floyd, P.A., Holder, M. and Leveridge, B., 1989 – The Rhenohercynian ocean revisited. Terra Abstracts 1 (1): 365 (abstract)

Frazier, W.J. and Schwimmer, D.R., 1987 – Regional stratigraphy of North America. Plenum Press, New York, London, 719 p

Friend, P.F., 1981 – Devonian Sedimentary Basins and deep faults of the northern most Atlantic Borderlands. Can. Soc. Petrol. Geol. Mem. 7: 149–165

Friend, P.F., Alexander-Marrack, P.E., Nicholsen, J. and Yeats, A.K., 1976 – Devonian sediments of East Greenland. I. Introduction, classification of sequences, petrographic notes, Medd. Grønland, 206 (1): 56 p.

Frisch, T., 1974 – Metamorphic and plutonic rocks and northern most Ellesmere Island, Canadian Arctic Archipellago. Geol. Surv. Canada Bull. 229

Fujita, K. and Cook, D.B., 1986 – Northeastern Siberia. In: The Arctic region, the decade of North American geology. Vol. 1, the Arctic ocean region Chapter G. continental margin (in press)

Fyffe, L.R. and Barr, S.M., 1986 – Petrochemistry and tectonic significance of Carboniferous volcanic rocks in New Brunswick. Can. J. Earth Sci., 23: 1243–1256

Fyffe, L.R., Rankin, D., Site, W.B. and Wones, D.R., 1982 – Volcanism and Plutonism in the Appalachian orogen. In: P.E. Schenk (ed.). Regional Trends in the Geology of the Appalachian-Caledonian-Hercynian-Mauretanic Orogen, D. Reidel Publ. Co., 173–185

Gabrielse, H., 1967 – Tectonic evolution of the northern Canadian Cordillera. Can. J. Earth Sci., 4: 271–298

Gardiner, P.R.R. and MacCarthy, I.A.J., 1981 – The late Palaeozoic evolution of southern Ireland in the context of tectonic basins and their transatlantic significance. Can. Soc. Petrol. Geol. Mem. 7: 683–725

Gayer, R.A. (ed.) , 1985 – The tectonic evolution of the Caledonide-Appalachian orogen. F. Vieweg & Son, Braunschweig/Wiesbaden, 194 p

Gee, D.G., and Sturt, B.A. (eds.), 1986 – The Caledonide Orogen – Scandinavia and related areas. John Wiley & Sons, part 1 and 2, 1266 p

Gilluly J., 1963 – The tectonic evolution of the western United States. Quart J. Geol. Soc. Lond., 119: 113–174

Glennie, K.W., 1986 – Early Permian-Rotliegend. In: K.W. Glennie (ed.). Introduction to the petroleum geology of the North Sea. Blackwell Scientific Publications Oxford, London, Edinburgh, 63–85

Gobbet, D.J., 1973 – Permian Fusulinacea. In: A. Hallam (ed.). Atlas of Palaeobiogeography. Elsevier Scientific Publ. Co., Amsterdam, London, New York, 151–158

Gordey, S.P., 1988 – Devono-Mississippian sedimentation and tectonism in the Canadian Cordilleran Miogeocline. In: N.J. McMillan, A.F. Embry and D.J. Glass (eds.). Devonian of the World. Can. Soc. Petrol. Geol. Mem.

14, Vol. 2:1–14

Gortunov, G.I., Zagorodny, V.G., Makievsky, S.I. and Suetnor, V.V., 1984 – The structure and history of the formation of the Barents Sea South Frame. In: Arctic Geology. Proc. 27th Internat. Geol. Congr., coll. 04, vol. 4: 84–93

Gramberg, I.S., Kos'ko, M.K. and Pogrebitskiy, Yv.Ye., 1986 – Tectonic evolution of the Arctic Shelf of Siberia from Riphean through Mesozoic time. Int. Geol. Review, 28 (8): 943–955

Grantz, A., Eittreim, S. and Whitney, O.T., 1981 – Geology and Physiography of the continental margin of Alaska and implications for the origin of the Canada Basin. In: A.E.M. Nairn, M. Churkin jr. and F.G. Stehli (eds.). The Ocean Basin and Margins, vol. 5, The Arctic Ocean, Plenum Press (New York, London): 439–492

Greiner, H., 1978 – Late Devonian facies interrelationships in bordering areas of the North-Atlantic and their palaeogeographic implications. Palaeogr., Palaeoclim., Palaeoecol., 25: 241–263

Guillocheau, F. and Rolet, J., 1982 – La sédimentation paléozoique Ouest-Armoricaine. Bull. Soc. géol. Mineral, Bretagne, (C) 14 (2): 45–62

Håkansson, E. and Pedersen, S.A.S., 1982 – Late Palaeozoic to Tertiary tectonic evolution of the continental margin in North Greenland. In: A.F. Embry and H.R. Balkwill (eds.). Arctic Geology and Geophysics, Can. Soc. Petrol. Geol. Mem. 8: 331–348

Håkansson, E. and Stemmerik, L., 1984 – Wandel Sea Basin, the North Greenland equivalent to Svalbard and the Barents shelf. In: A.M. Spencer et al. (eds.). Petroleum geology of the north European margin. Norw. Petrol. Soc., Graham & Trotman, London, 97–107

Hälbich, I.W., 1983 – Tectogenesis of the Cape Fold Belt (CFB). In: A.P.G. Söhnge and

I.W. Hälbich (eds.). Geodynamics of the Cape Fold Belt. The Geological Society of South Africa, Spec. Publication No. 12: 165–175.

Haller, J., 1971 – Geology of the East Greenland Caledonides. In: L.V. de Sitter (ed.). Regional Geology Series. Interscience Publ., London, 413 p

Hambrey, M.J. and Harland, W.B. (eds.), 1981 – Earth's pre-Pleistocene glacial record. Cambridge University Press, 1044 p

Hamilton, R.F.M. and Trewin, N.H., 1988 – Environmental controls on fish faunas of the Middle Devonian Orcadian Basin. In: N.J. McMillan, A.F. Embry and D.J. Glass (eds.). Devonian of the World. Can. Soc. Petrol. Geol. Mem. 14, Vol. 3: 584–600

Handschy, J.W. and Dyer, R., 1987 – Polyphase deformation in Sierra del Cuervo, Chihuahua, Mexico: evidence for Ancestral Rocky Mountain tectonics in the Ouachita foreland of northern Mexico. Geol. Soc. Am. Bull. 99: 618–632

Handschy, J.W., Keller, G.R. and Smith, K.J., 1987 – The Ouachita system in northern Mexico. Tectonics 6 (3): 323–330

Hargraves, R.B., Dawson, E.M. and Van Houten, F.B., 1987 – Paleomagnetism and age of Mid-Paleozoic Ring Complexes in Niger, West Africa and tectonic implications. Geophys. J. Roy. astr. Soc., Vol.90(3):705–729.

Harland, W.B., Cox, A.V., Llewellyn, P.G., Pickton, C.A.G., Smith, A.G. and Walters, R., 1982 – A geologic time scale. Cambridge Earth Science Series, Cambridge University Press, 131 p

Harland, W.B., Gaskell, B.A., Heafford, A.P., Lind, E.K. and Perkins, P.J., 1984 – Outline of Arctic post-Silurian continental displacements. In: A.M. Spencer et al. (eds.). Petroleum geology of the North European margin. Norw. Petrol. Soc.; Graham & Trotman Ltd., London: 137–148

Hatcher, R.D., jr., 1981 – Thrusts and nappes in the North American Appalachian Orogen. In: K.R. McClay and N.J. Price (eds.). Thrust and nappe tectonics. Geol. Soc. Lond., Spec. Publ. 9: 491–499

Hatcher, R.D., jr., 1985 – The North American Appalachian Orogen. In: R.A. Gayer (ed.). The tectonic evolution of the Caledonide-Appalachian orogen, F. Vieweg & Son, Braunschweig/Wiesbaden, 48–56

Hatcher, R.D., jr., 1987 – Tectonics of the southern Central Appalachian internides. Ann. Rev. Earth Planet. Sci. 15: 337–362

Heafford, A.P., 1988 – Carboniferous through Triassic stratigraphy of the Barents shelf. In: W.B. Harland and E.K. Dowdeswell (eds.). Geological evolution of the Barents shelf region. Graham & Trotman, London, Dordrecht, Boston, 89–109

Heller, P.L. and Angevine, C.L., 1985 – Sea-level cycles during the growth of Atlantic-type oceans. Earth Planet. Sci. Lett. 75: 417–426

Henes-Klaiber, U., Holl, A. and Altherr, R., 1989 – The Odenwald: more evidence for Hercynian arc magmatism Terra Abstracts 1 (1): 281 (abstract)

Henriksen, N. and Higgins, A.K., 1976 – East Greenland Caledonian fold belt. In: A. Escher and W.S. Watt (eds.). Geology of Greenland. Greenland Geol. Surv.: 183–246

Hizhnyakov, A.V. and Pomyanovskaya, G.M., 1967 – Devonian of the Volyno-Podolian margin of the Russian platform. In: D.H. Oswald (ed.). International Symposium on the Devonian System Calgary, Alberta. Soc. Petrol. Geol., vol. I: 359–367

Holder, M.T., Leveridge, B.E., 1986 – A model for the tectonic evolution of South Cornwall. J. Geol. Soc. Lond., 143: 125–134

Holl, A. and Altherr, R., 1987 – Hercynian I-type granitoids of northern Vosges: documents of increasing arc maturity. Terra Cognita, 7 (2–3): 74 (abstract)

Holtedahl, O., 1960 – Geology of Norway. Norges Geol. Undersokelse, 208, 540 p.

Hossack, J.R., 1985 – The role of thrusting in the Scandinavian Caledonides. In: R.A. Gayer (ed.). The tectonic evolution of the Caldonide-Appalachian orogen, F. Vieweg & Sohn, Braunschweig/Wiesbaden: 97–116

Hossack, J.R., Garton, M.R. and Nickelsen, R.P., 1986 – The geological section from the foreland up to the Jotun thrust sheet in the Valderes area, South Norway. In: D.G. Gee and A.B. Sturt (eds.). The Caledonide Orogen-Scandinavia and related areas. John Wiley & Sons Ltd.: 443–456

House, M.R., 1975 – Facies and time in Devonian tropical areas. Proceedings of the Yorkshire Geol. Soc., 40 (2) 16: 233–288

House, M.R., 1983 – Devonian eustatic events. Proc. Ussher Soc., 5: 396–405 (Annual Conference of the Ussher Society Jan. 1983)

House, M.R., Richardson, J.B., Chaloner, W.G., Alten, J.R.L., Holland, C.H. and Westoll, T.S., 1977 – A correlation of Devonian rocks of the British Isles. Geol. Soc. Lond., Spec. Report 8: 110 p

Howell, D.G. and Wiley, T.J., 1987 – Crustal evolution of northern Alaska inferred from sedimentology and structural relations of the Kandik area. Tectonics, 6 (5): 619–631

Howie, R.D. and Barss, M.S., 1975 – Palaeogeography and sedimentation in the Upper Palaeozoic, eastern Canada. In: C.J. Yorath, E.R. Parker and D.J. Glass (eds.). Canada's continental margins and offshore petroleum exploration. Can. Soc. Petrol. Geol. Mem. 4: 45–57

Hubacher F.A. and Lux D.R., 1987 – Timing of Acadian deformation in northeastern Maine. Geology, 15: 80–83

Hubbard, R.J., Edrich, S.P. and Rattey, R.P., 1987 – Geologic evolution and hydrocarbon habitat of the "Arctic Alaska microplate". Mar. Petrol. Geol., 4 (1): 2–34

Hurst, J.M., McKerrow, W.S., Soper, N.J. and Surlyk, F., 1983 – The relationship between Caldonian nappe tectonics and Silurian turbidite deposits in Northern Greenland. J. Geol. Soc. Lond., 140: 123–131

Hurst, J.M. and Surlyk, F., 1984 – Tectonic control of Silurian Carbonate shelf margin morphology and facies, North Greenland. Am. Ass. Petrol. Geol. Bull. 68 (1): 1–17

Hutton, D.H.W. and Murphy, F.C., 1987 – The Silurian of the southern Uplands and Ireland as a successor basin to the end-Ordovician closure of Iapetus. J. Geol. Soc. Lond., 144 (5): 765–772

Iglesias, M., Ribeiro, M.L. and Ribeiro, A., 1983 – La interpretacion aloctonista de la estrutura del noroeste Penisular. In: Libro Jubilar J.M. Rios, Geologia de Espana, Tomo I., Inst. Geol., Minero Espana, Madrid, Publ.: 459–467

Igolkina, N.S. and Krivskaya, T.Y., 1978 – The pre-Visean hiatus on the Russian Platform. International Geology Review 20(4): 407–412

Isaac, K.P. and Barnes, R.P., 1985 – Discussion of papers on the Hercynian back-arc marginal basin of SW England. J. Geol. Soc. Lond., 142 (5): 927–929

Jarvis, G.T. and McKenzie, D.P., 1980 – Sedimentary basin formation with finite extension rates. Earth Planet. Sci. Lett. 48: 42–52

Johnson, G.A.L., 1981 – Geographical evolution from Laurasia to Pangea. Proc. Yorkshire Geol. Soc., 43 (3) 13: 221–252

Johnson, H.D., Levell, B.K. and Siedlicki, S., 1978 – Late Precambrian sedimentary rocks in East Finnmark, north Norway and their relationship to the Trollfjord-Komagelv fault. J. Geol. Soc. Lond., 135: 517–533

Johnson, J.G., Klapper, G. and Sandberg, C.A., 1985 – Devonian eustatic fluctuations in Euramerica. Geol. Soc. Am. Bull. 96: 567–587

Jones, M., van der Voo, R. and Bonhommet, N., 1979 – Late Devonian to Early Carboniferous palaeomagnetic poles from the Armorican Massif, France. Geophys. J.R. Astr. Soc. 58: 287–308

Juana, C.G. de, Iturralde de Arozona, J.M. and Cadillat, X.P., 1980 – Geologia de Venezuela y de sus Cuencas Petroliferas; Tomo I, Ediciones Foninues, Caracas. 407 p

Julivert, M., 1978 – Hercynian Orogeny and Carboniferous palaeogeography in northwestern Spain: a model of deformation-sedimentation relationship. Z. dt. geol. Ges., 129: 565–592

Julivert, M., 1979 – A cross-section through the northern part of the Iberian Massif, its position within the Hercynian fold belt. Krystallinikum, 14: 51–57

Julivert, M., 1983 – La evolucion sedimentaria durante el Paleozoico y el registro de la deformacion en la columna estratigraphica paleozoica. In: Libro Jubilar J.M. Ramos, Geologica de Espana, Tomo I., Inst. Geol. Minero Espana, Madrid Publ., 593–600

Julivert, M., Marcos, A. and Pérez-Estaún, A., 1977 – La structure de la chaîne Hercynienne dans le secteur Ibérique et l'arc Ibéro-- Armoricain. In: La chaîne varisque d'Europe moyenne et occidentale, Colloque Int. CNRS, Rennes, 243: 429–440

Julivert, M., Truyols, J. and Vergés J., 1983 – El Devonico en el Macizo Iberico. In: Libro Jubilar J.M. Ramos, Geologia de Espana, Tomo I, Inst. Geol. Minero Espana, Madrid Publ.: 265–310

Kamen-Kaye, M., 1976 – Mediterranean Permian Tethys. Bull. Am. Assoc. Petrol. Geol., 60: 623–626

Kazantseva, T.T. and Kamaletditov, M.A., 1986 – The geosynclinal development of the Urals. Tectonophysics 127: 371–381

Kent, D.V. and Keppie, J.D., 1988 – Silurian-Permian paleocontinental reconstructions and circum-Atlantic tectonics. In: A.L. Harris and D.J. Fettes (eds.). The Caledonian-Appalachian Orogen. Geol. Soc. Lond., Spec. Publ.: 469–480

Keppie, J.D., 1982 – Tectonic map of the Province of Nova Scotia, 1:500.000. Dept. of Mines and Energy, Nova Scotia

Keppie, J.D., 1985 – Geology and tectonics of Nova Scotia. In: Appalachian geotraverse (Canadian mainland), field excursions. Vol. 1 Geol. Ass. of Canada, Mineral Ass. Can., University of New Brunswick, 23–108

Keppie, J.D. and Dallmeyer, R.D., 1987 – Dating transcurrent terrane accretion: an example from the Meguma and Avalon composite terranes in the northern Appalachians. Tectonics 6 (6): 831–847

Keppie, J.D., Currie, K.L., Murphy, J.B., Pickerill, R.K., Fyffe, L.R. and St. Julien, P., 1985 – Appalachian geotraverse (Canadian mainland), field excursions. Vol. I Geol. Ass. Canada, Mineral Ass. Can., University of New Brunswick, 181

Kerr, J.Wm., 1967 – The Franklinian Miogeosyncline and adjacent central stable region, Arctic Canada. In: D.H. Oswald (ed.). International Symposium on the Devonian System, Alberta. Soc. Petrol. Geol. Vol. 1: 677–692

Kerr, J.Wm., 1977 – Cornwallis Fold Belt and the mechanism of basement uplift. Can. J. Earth Sci., 14 (6): 1374–1401

Kerr, J.Wm., 1981 – Evolution of the Canadian Arctic Islands: a transition between the Atlantic and Arctic Ocean. In: A.E.M. Nairn, M. Churkin jr. and F.G. Stehli (eds.). The Ocean Basins and Margins, vol. 5, the Arctic Ocean, Plenum Press, New-York, 105–199

Khain, V.E., 1984 – The Alpine-Mediterranean fold belt of the USSR. Episodes, 7 (3): 20–29

Khain, V.E., 1985 – Geology of the USSR, first part: Old Cratons and Paleozoic fold belts. Beiträge zur Regionalen Geologie der Erde, Band 17, Gebrüder Bornträger, Berlin, Stuttgart, 272 p.

Kirton, S.R., 1984 – Carboniferous volcanicity in England with special reference to the Westphalian of the E. and W. Midlands. J. Geol. Soc. Lond., 141 (1): 161–170

Klein, G. de V. and Hsui, A.F., 1987 – Origin of cratonic basins. Geology 15: 1094–1098

Klein, R.J. and Barr, M.V., 1986 – Regional state of stress in Western Europe. In: O. Stephenson (eds.). Rock stress and rock stress measurements. Centrale Publishers, Lund, Sweden, 33–44

Kluth, C.F., 1986 – Plate tectonics and the Ancestral Rocky Mountains. In: J.A. Peterson (ed.). Paleotectonics and sedimentation in the Rocky Mountain area, United States. Am. Ass. Petrol. Geol. Mem. 41: 353–369

Kulander, B.R. and Dean, S.L., 1986 – Structure and tectonics of central and southern Appalachian Valley and ridge and plateau provinces, West Virginia and Virginia. Am. Assoc. Petrol. Geol. Bull. 10 (11): 1624–1684

Kullmann, J., Schönenberg, R. and Wiedmann, J. (eds.), 1982 – Subsidenz-Entwicklung im Kantabrischen Variszikum und an passiven Kontinentalrändern der Kreide, Teil 1, Variszikum. N. Jb. Geol. Palont., 163 (2): 137–300

Lamar, D.L, Reed, W.E. and Douglass, D.N., 1986 – Billefjorden fault zone, Spitsbergen: Is it part of a major Late Devonian transform? Geol. Soc. Am. Bull., 97: 1083–1088

Lardeux, H., Chauvel, J.J., Hery, J.L., Morzadec, P., Paris, F.L., Racheboeuf, P. and Robardet, M., 1977 – Evolution géologique du Massif Armoricain au cours des temps Ordoviceins, Siluriens et Devoniens. In: La chaine varisque d'Europe moyenne et occidentale. Coll. intern. CNRS, Rennes, 243: 181–192

Lécorché, J.P., 1983 – Structure of the Mauretanides. In: P.E. Schenk (ed.). Regional trends in the geology of the Appalachian-Caledonian-Hercynian-Mauretanide orogen. D. Reidel Publ. Co.: 347–353

Lécorché, J.P. and Clauer, N., 1983 – First radiometric data (K/Ar) on the front of Mauretanides in the Akjouit region (Mauretania). In: The Caledonide orogen, IGCP project 27, Symposium, Morocco and Paleozoic orogenesis, Rabat, p. 23

Lécorché, J.P. and Sougy, J., 1978 – Les Mauretanides, Afrique Occidentale, Essai de synthèse. In: Caledonian-Appalachian orogen of the North Atlantic region. Geological Survey of Canada, paper 78–14: 231–242.

Leeder, M.R., 1976 – Sedimentary facies and the origin of basin subsidence along the northern margin of the supposed Hercynian Ocean. Tectonophysics, 36: 167–179

Leeder, M.R., 1982 – Upper Palaeozoic basins in the British Isle – Caledonian inheritance verus Hercynian plate margin processes. J. Geol. Soc. Lond., 139: 479–491

Leggett, J.K., McKerrow, W.S. and Soper, N.J., 1983 – A model for the Curstal Evolution of Southern Scotland. Tectonics, 2 (2): 187–210

Lethiers, F., Braun, W.K., Crasquin, S. and Mansy, J.-C., 1986 – The Strunian event in western Canada with reference to ostracod assemblages. Ann. Soc. géol. Belgique, 109: 149–157

Lewandowski, M., 1987 – Results of preliminary paleomagnetic investigations of some lower Paleozoic rocks from the Holy Cross Mountains (Poland). Kwartalnik Geologiczny, 31(4): 543–556

Lorenz, V. and Nicholls, J.A., 1984 – Plate and intra-plate processes of Hercynian Europe during the Late Palaeozoic. Tectonophysics, 107: 25–56

Lützner, H. and Schwab, G., 1982 – Tectonic regime of Molasse Epochs. Akad. Wiss. DDR Veröff. Zent. Inst. Physik der Erde, 66: 395 p. and enclosures

Mack, G.H., Thomas, W.A. and Horsey, C.A., 1983 – Composition of Carboniferous sandstones and tectonic framework of southern Appalachian-Ouachita orogen. J. Sed. Petrology, 53, (3): 931–946

Manspeizer, W. (ed.) 1988 – Triassic-Jurassic rifting, continental break-up and the Atlantic Ocean and passive margins. Vol. A and B.

Developments in Geotectonics 22. Elsevier, Amsterdam, Oxford, New York, Tokyo, 998 p

Martinez-Garcia, E., 1972 – El Silurico de San Vitero (Zamora), Comparacion con series vecinas e importancia orogenica. Acta geol. Hispanica, VII (4): 104–108

Martinez-Garcia, E., 1980 – El Macizo Herciniano del dominio Iberico Central. Bol. Geol. Minero, XCI-I: 67–90

Matte, Ph., 1983 – Two geotraverses across the Ibero-Armorican Variscan arc of Western Europe. In: Profiles of orogenic belts. Geodyn. Ser. 20, Am. Geophys. Union, Geol. Soc. Am., 53–81

Matte, Ph., 1986 – La chaîne varisque parmi les chaînes paléozoiques peri-atlantiques, modèle d'évolution et position des grandes blocs continenteaux au Permo-Carbonifère. Bull. Soc. géol. France (8) II (1): 9–24

Matviyevskaya, N.O., Bogatsky, V.I., Vasserman, B.Ya., Gobanov, L.A. and Shafran, Ye.B., 1986 – Geological history of oil-gas potential of the Pechora-Kolvin aulacogen. Petroleum Geology 21 (7): 301–305

McKee, E.D. and Crosby, E.J., 1975 – Paleotectonic investigations of the Pennsylvian system in the United States. Geol. Surv. Professional Paper 853 United States Government Printing Office, Washington. Vol. 1, 349 p., vol. 2, 192 p., vol. 3, plates

McKenzie, D.P. 1969 – Speculations on the consequences and causes of plate motions. Geol. J. Roy. Astron. Soc. 18: 1–32

McKenzie, D.P., 1978 – Some remarks on the development of sedimentary basins. Earth Planet. Sci. Lett. 40: 25–32

McKerrow, W.S., 1988 – Wenlockian to Givetian deformation in the British Isles and the Canadian Appalachians. In: A.L. Harris and D.J. Fettes (eds.). The Caledonian-Appalachian Orogen. Geol. Soc. Spec. Publ. 38: 437–448

Meissner, R., Bartelsen, H. and Murawski, H., 1981 – Thin skinned tectonics in the north-ern Rhenish Massif, Germany. Nature 290: 399–401

Mesner, J.L. and Woodridge, L.C.P., 1964 – Maranhao Paleozoic basin and Cretaceous coastal basins, North Brazil. Bull. Am. Assoc. Petrol. Geol. 48 (9): 1475–1512

Metz, P.A., Egan, A. and Johansen, O., 1982 – Landsat linear features and incipient Rift system model for the origin of base metal and petroleum resources of Northern Alaska. In: A.F. Embry and H.R. Balkwill (eds.). Arctic Geology and Geophysics. Can. Soc. Petrol. Geol. Mem. 8: 101–112

Meyer, W. and Stets, J., 1980 – Zur Paläogeographie von Unter und Mitteldevon im westlichen und zentralen Rheinischen Schiefergebirge. Z. dt. geol. Ges., 131: 725–751

Michard, A., Yazidi, A., Benziane, F., Hollard, H. and Willefert, S., 1982 – Foreland thrusts and olistromes on the pre-Sahara margin of the Variscan orogen, Morocco. Geology, 10: 253–256

Miller, J.D. and Kent, D.V., 1986 – Paleomagnetism of the Upper Devonian Catskill Formation from the southern limb of the Pennsylvania salient. Geophys. Research Letters, 13: 1173–1176

Miller, M.M., 1987 – Dispersed remnants of a northeast Pacific fringing arc: Upper Paleozoic terranes of Permian McCloud faunal affinity, western US. Tectonics, 6: 807–830

Mitchell, A.H.G., 1981 – The Grampian Orogeny in Scotland and Ireland: almost an ancient Taiwan. Proc. Geol. Soc. China, 24: 113–129

Moore, T.E. and Nilsen, T.H., 1984 – Regional variations in the fluvial Upper Devonian and Lower Mississippian Kanayut conglomerate, Brooks Range, Alaska. In: T.H. Nilsen (ed.). Fluvial sedimentation and related Tectonic Framework, Western North America. Sedimentary Geology, 38: 465–497

Morel, P. and Irving, E., 1978 – Tentative

paleocontinental maps for the Early Phanerozoic and Proterozoic. J. Geol., 86: 535–561

Morrow, D.W. and Geldsetzer, H.H.J., 1988 – Devonian of the Eastern Canadian Cordillera. In: N.J. McMillan, A.F. Embry and D.J. Glass (eds.). Devonian of the World. Can. Soc. Petrol. Geol. Mem. 14, Vol. 1: 85–122

Mortensen, J.K., 1982 – Geological setting and tectonic significance of Mississippian felsic metavolcanic rocks in the Pelly Mountains, southeastern Yukon territory. Can. J. Earth Sci., 19: 8–22

Mosher, S. and Rast, N., 1984 – The deformation and metamorphism of Carboniferous rocks in Maritime Canada and New England. In: D.H.W. Hutton and D.J. Sanderson (eds.). Variscan tectonics of the North Atlantic region. Geol. Soc. Lond. Spec. Publ. No. 14: 233–243

Nalivkin, V.D., 1982 – La fosse bordière de pré-Oural. In: Tectonics of Europe and adjacent areas, Variscides, Epi-Paleozoic platforms, Alpides. Explanatory note to the International Tectonic map of Europe and adjacent areas. Scale 1:2,500,000, "Nauka" Publ. House, Moscow: 206–213

Nance, R.D., 1987 – Dextral transpression and Late Carboniferous sedimentation in the Fundy coastal zone of southern New Brunswick. In: C. Beaumont and A.J. Tankard (eds.). Sedimentary basins and basin forming mechanisms. Can. Soc. Petrol. Geol. Mem. 12: 363–377

Nilsen, T.H., 1973 – Devonian (Old Red) Sedimentation and Tectonics of Norway. In: M.G. Pitcher (ed.). Arctic Geology. Am. Assoc. Petrol. Geol. Mem. 19: 471–481

Nilsen, T.H., 1981 – Upper Devonian and Lower Mississippian Redbeds, Brooks Range, Alaska. In: A.D. Miall (ed.). Sedimentation and Tectonics in Alluvial Basins. Geol. Ass. Canada, Special Paper 23: 187–219

Nilsen, T.H., 1983 – Influence of the Greenland-Scotland ridge on the geological history of the North Atlantic and Norwegian-Greenland Sea areas. In: H.P. Bott, S. Saxov, M. Talwani and J. Thiede (eds.). Structure and development of the Greenland-Scotland Ridge – new methods and concepts. Plenum Press (New York, London), 457–478

Nilsen, T.H. and Stewart, J.H., 1980 – The Antler orogeny – Mid-Paleozoic tectonism in western North America. Geology, 8: 298–302

Norton, M.G., 1986 – Late Caledonide extention in Western Norway. Tectonics, 5 (2): 195–204

Oliveira, J.T., 1982 – The Devonian-Carboniferous stratigraphy and geodynamics of southern Portugal: some comments. N. Jb. Geol. Palaeont., 163 (2): 276–284

Oliveira, J.T., Garcia-Alcade, J.C., Linan,E. and Truyols, J., 1986 – The Famennian of the Iberian Peninsula. Ann. Soc. géol. Belgique, 109: 159–174

Oliver, W.A. jr., De Witt, W. jr., Dennison, J.M., Hoskins, D.M. and Huddle, J.W., 1967 – Devonian of the Appalachian Basin, United States. In: D.H. Oswald (ed.). International Symposium on the Devonian System, Alberta. Soc. Petrol. Geol., vol. 1: 1001–1040

Opdyke, N.D., Jones, D.S., MacFadden, B.J., Smith, D.L., Mueller, P.A. and Shuster, R.D., 1987 – Florida as an exotic terrane: Paleomagnetic and geochronologic investigation of Lower Paleozoic rocks from the subsurface of Florida. Geology, vol 15: 900–903

Osberg, P.H., 1982 – Timing of orogenic events to the U.S. Appalachians. In: P.E. Schenk (ed.). Regional trends in the geology of the Appalachian-Caledonian-Hercynian-Mauretanide orogen. D. Reidel Publ. Co., 315–337

Oswald, D.H. (ed.), 1967 – International Symposium on the Devonian Syste, Calgary 1967. Alberta Soc. Petrol. Geol., vol. I and II, 1377 p

Pamic, J.J., 1984 – Triassic magmatism of the Dinarides in Yugoslavia. Tectonophysics, 109: 273–307

Paproth, E., 1978 – Nicht-marine Muscheln als Spiegel der Fazies Entwicklung im paralischen Kohlengebiet Northwest-Europas. Sonderveröff. Geolog. Inst. Univ. Köln, No. 33: 91–100

Paproth, E., Dreesen, R. and Thorez, J., 1986 – Famennian paleogeography and event stratigraphy of Northwest Europe. Ann. Soc. géol. Belgique, 109: 175–186

Paris F., Morzadec P., Herisse A. and Pelhate A., 1986 – Late Devonian-Early Carboniferous events in the Armorican Massif (western France): a review. Ann. Soc. géol. Belgique, 109: 187–195

Park, R.G., 1988 – Geological structures and moving plates. Blackie, Glasgow, London, 337 p

Parrish, J.M., Totman-Parrish, J. and Ziegler, A.M., 1986 – Permian-Triassic paleogeography and paleoclimatology and implications for Tetrapsid distribution. In: N. Hotton III, P.D. MacLean, J.J. Roth and E.C. Roth (eds.). The ecology and biology of Mammal like reptiles. Smithonian Institution Press, Washington and London, 109–131

Pavoni,N., 1985 – Die pazifisch-antipazifische Bipolarität im Strukturbild der Erde und ihre geodynamische Deutung. Geol. Rundschau, 74: 251–266

Pavoni, N., 1988 – Das bipolare geotektonische Modell. Ein Versuch zur Beschreibung der grossräumlichen Bewegungsvorgänge im Erdmantel. Geographische Rundschau, 40 (10): 58–64

Pedersen, S.A.S. and Holm, P.M., 1983 – The significance of a Middle Devonian K/Ar age of an intrusive rock in the southern part of the North Greenland Belt. Bull. geol. Soc.

Denmark, 31: 121–127

Perch-Nielsen, K., Bromley, R., Birkenmajer, K. and Allen, M., 1972 – Field observation in Palaeozoic and Mesozoic sediments of Scoresby Land and northern Jameson Land. Rapp. Grønlands geol. Unders., 48: 39–59

Pérez-Estaún, A., Bastida, F., Alonso, J.L., Marquinez, J., Aller, J., Alvarez-Marrón, J., Marcos, A. and Pulgar, J.A., 1988 – A thin skinned tectonic model for an arcuate fold and thrust belt: the Cantabrian zone (Variscan Ibero-Armorican arc). Tectonics 7 (3): 517–537

Peterson J.A., 1986 – General Stratigraphy and Regional Paleotectonics of the Western Montana Overthrust Belt. In: J.A. Peterson (ed.). Paleotectonics and sedimentation in the Rocky Mountain Region, United States, Am. Assoc. Petrol. Geol. Mem.: 41: 51–86

Peterson, J.A. and Smith, D.L., 1986 – Rocky Mountain paleogeography through geologic time. In: J.A. Peterson (ed.). Paleotectonics and sedimentation in the Rocky Mountain region, United States. Am. Assoc. Petrol. Geol. Mem. 41: 3–19

Pfeiffer, H., 1971 – Die variszische Hauptbewegung (sogenannte Sudetische Phase) im Umkreis des äusseren Kristallinzone des variszischen Bogens. Geologie, 20: 945–958

Phillips, W.E.A., Stillmann, Ch.J. and Murphy, T., 1976 – A Caledonian plate tectonic model. J. Geol. Soc. London, 132 (6): 579–609

Pindell, J.L., 1985 – Alleghenian reconstruction and subsequent evolution of the Gulf of Mexico, Bahamas and Proto-Caribbean. Tectonics, 4 (1): 1–39

Pindell, J. and Dewey, J.F., 1982 – Permo-Triassic reconstruction of western Pangea and the evolution of the Gulf of Mexico/Caribbean region. Tectonics, 1 (2): 179–211

Piqué, A., 1981 – Un segment de chaîne intracontinentale: la Meseta marocaine nord-occidentale, influence des fractures du socle précambrien sur la sédimentation et la déformation de la couverture Palozoique. Bull. Soc. géol. France, (7) XXIII (1): 3–10

Piqué, A., 1982 – Structural Domains of the Hercynian Belt in Morocco. In: P.E. Schenk (ed.). Regional Trends in the Geology of the Appalachian-Caledonian-Hercynian-Mauretanide Orogen. D. Reidel Publ. Co.: 339–345

Piqué, A. and Kharbouch, F., 1983 – Distension intra continentale et volcanisme associé, la Meseta Marocaine nord-occidentale au Devonien-Dinantien. Bull. Centres Rech. Explor.-Prod. Elf-Aquitaine, 7 (1): 377–387

Platen, K., Emmermann, R. and Franke, W., 1989 – Devonian MORB-type metabasalts from the Rhenohercynian zone in Germany. Terra Abstracts 1 (1): 365 (abstract)

Pluijm, B. van der, 1987 – Timing and spatial distribution of deformation in the Newfoundland Appalachians: a "multi-stage collision" history. Tectonophysics, 135: 15–24

Pogrebitsky, Y.E., 1982 – Le système plissé du Taimyr. In: A.V. Peive, K.E. Khain, M.V. Mouratov and F. Delany (eds.). Tectonic of Europe and adjacent areas. Nauka Publishing House, Moscow, 229–230

Poole, F.G., Baars, D.L., Drewes, H., Haykes, P.T., Ketner, K.B., McKee, E.D., Teichert, C. and Williams, J.S., 1967 – Devonian of the Southwestern United States. In: D.H. Oswald (ed.). International Symposium on the Devonian System, Calgary, Alberta. Soc. Petrol. Geol., vol. I: 879–913

Poole, W.H., 1977 – Evolution of the Appalachian belt in Canada. In: La Chaîne varisque d'Europe moyenne et occidentale, Colloque internat. CRNS Rennes, 243: 587–603

Poole, W.H., Sanford, B.V., Williams, A. and Kelley, D.G., 1970 – Geology of Southeastern Canada. In: Geology and Economic Minerals of Canada, Economic Geology Report no. 1, 5th edition Dept. of Energy, Mines and Resources, Ottawa, Canada: 227–304

Powell, D., Andersen, T.B., Drake, A.A., Hall, L. and Keppie, J.D., 1988 – The age and distribution of basement rocks in the Caledonide orogen of the N. Atlantic. In: A.L. Harris and D.J. Feltes (eds.). The Caledonian-Appalachian Orogen. Geol. Soc., Lond. Spec. Publ. 38: 63–74

Quinlan, G., 1987 – Models of subsidence mechanisms in intracratonic Basins, and their applicability to North American examples. In: C. Beaumont and A.J. Tankard (eds.). Sedimentary basins and basin-forming mechanisms. Can. Soc. Petrol. Geol. Mem.12: 463–481

Quinlan, G.M. and Beaumont, C., 1984 – Appalachian thrusting, lithospheric flexure and the Paleozoic stratigraphy of the eastern interior of North America. Can. J. Earth Sci., 21: 973–996

Raoult, J.F., 1986 – Le front Varisque du Nord de la France d'après les profils sismiques, la géologie de surface et les sondages. Rev. Géol. Dynam. Georgr. Phys., 27 (3–4): 247–268

Rast, N., 1983 – The northern Appalachians traverse in maritime Canada. In: N. Rast and F.M. Delany (eds.). Profiles of orogenic belts, Geodynamic Series, vol. 10, Am. Geophys. Union; Geol. Soc. Am., Washington, 243–274

Rast, N., 1984 – The Alleghenian orogeny in eastern North America. In: D.H.W. Hutton and D.J. Sanderson (eds.). Variscan tectonics of the north Atlantic region. Geol. Soc. Lond. Spec. Publ. 14: 197–218

Rayer, F.G., 1981 – Exploration prospect and future petroleum potential of the Canadian Arctic Islands. J. Petrol. Geol. 3 (4): 367–412

Raymond, A., 1987 – Paleogeographic distribution of Early Devonian plant traits. Palaios, 2: 113–132

Rex, D.C. and Higgins, A.K., 1985 –

Potassium-argon mineral ages from the East Greenland Caledonides between 72° and 74 °N. In: D.G. Gee and B.A. Sturt (eds.). The Caledonide Orogen-Scandinaivia and Related Areas. John Wiley & Sons. Part 2: 1115–1125

Reynolds, P.H., Zenthill and Mcke, G.K., 1981 – K-Ar and 40Ar/39Ar geochronology of granitoid rocks from southern Nova Scotia: its bearing on the evolution of the Meguma Zone of the Appalachians. Can. J. Earth Sci., 18: 386–394

Reynolds, P.H., Elias, P., Muecke, G.K. and Grist, A.M., 1987 – Thermal history of the southwestern Meguma zone, Nova Scotia, from 40Ar/39Ar and fisson track dating study of intrusive rocks. Can. J. Earth Sci. 24: 1952–1965

Roberts, D., 1983 – Devonian tectonic deformation in the Norwegian Caledonides and its regional perspective. Norges Geol. Unders., 380: 85–96

Roberts, D., 1988 – Timing of Silurian to Middle Devonian deformation in the Caledonides of Scandinavia, Svalbard and E. Greenland. In A.L. Harris and D.J. Fettes (eds.). The Caledonian-Appalachian Orogen. Geol. Soc., Lond. Spec. Publ. 38: 429–435

Roberts, D. and Gale, E.H., 1978 – The Caledonian-Appalachian Iapetus Ocean. In: D.H. Tarling (ed.). Evolution of the Earth's Crust, Academic Press, London, 255–342

Roddick, J.C., Cameron, W.E. and Smith A.G., 1979 – Permo-Triassic and Jurassic 40Ar/39Ar ages from Greek ophiolites and associated rocks. Nature, Lond., 279: 788–790

Roeder, E., Gilbert, D.E., jr. and Witherspoon, W.D., 1978 – Evolution and macroscopic structure of the Valley and Ridge thrust belt Tennesee and Virgina. Univ. Tenn. geol. Sci. Studies in Geology, 2. Knoxville, 25 p

Rolet, J., 1983 – La "phase Bretonne" en Bretagne: état des connaissances. Bull. Soc. géol. mineral, Bretagne, 1982 (C) 14 (2): 63–71

Rolet, J., Le Gall, B., Darboux, J.R., Thonon, P. and Gravelle, M., 1986 – L'évolution géodynamique dévono-carbonifère de l'extrémité occidentale de la chaîne hercynienne d'Europe sur le transect Armorique-Cornwall. Bull. Soc. géol. France, (8) II (1): 43–54

Romanovitch, B.S., 1982 – Archipel de Novaia Zemlia (ex-Nouvelle Zemble) et de l'ile Vaigatsch. In: A.V. Peive, V.E. Khain, M.V. Muratov and F. Delany (eds.). Tectonics of Europe and adjacent areas, Varisciden, Epi-Paleozoic platforms; explanatory note to the international tectonic map of Europe and adjacent areas. Scale 1:2,500,000: 203–206

Romer, A.S., 1973 – Permian reptiles. In: A. Hallam (ed.). Atlas of Paleobiogeography. Elsevier Scientific Publ. Co., Amsterdam, London, New York, 159–167

Rønnevik, H. Ch. and Beskow, B., 1983 – Structural and Stratigraphic Evolution of the Barents Sea. Norwegian Petroleum Directorate, Contribution No. 1, 29 p

Rønnevik, H. and Jacobsen, H.P., 1984 – Structural Highs and Basins in the Western Barents Sea. In: A.M. Spencer et al. (eds.). Petroleum Geology of the North European Margin. Norwegian Petrol. Soc., Graham & Trotman Ltd.,: 19–32

Ronov, A., Khain, V. and Seslavinsky, K., 1984 – Atlas of lithological-palaeogeographical maps of the world. Late Precambrian and Paleozoic of continents. USSR Academy of Science, Leningrad, 70 p

Ross, C.A., 1973 – Carboniferous foraminifera. In: A. Hallam (ed.). Atlas of Paleobiogerapy. Elsevier Scientific Publ. Co., Amsterdam, London, New York, 127–132

Ross, C.A. and Ross, J.R.P., 1985a – Paleozoic tectonics and sedimentation in West Texas, southern New Mexico and southern Arizona. In: P.W. Dickerson and W.R. Muehlberger (eds.). Structure and tectonics of Trans-

Pecos Texas. West Texas Geological Society Field Conferences, Publication 85–81: 221–230

Ross, C.A. and Ross, J.R.P., 1985b – Carboniferous and Early Permian biogeography. Geology, 13: 27–30

Ross, C.A. and Ross, J.R.P., 1986 – Paleozoic paleotectonics and sedimentation in Arizona and New Mexico. In: J.A. Peterson (ed.). Paleotectonics and sedimentation in the Rocky Mountain region, United States. Am. Assoc. Petrol. Geol. Mem. 41: 653–668

Ross, C.A. and Ross, J.R.P., 1987 – Late Paleozoic sea levels and depositional sequences. Cushman Foundation for Foraminiferal Research, Spec. Publ. 24: 137–149

Roussel J., Lécorché, J.P., Ponsard, J.F., Sougy, J. and Villeneuve M., 1984 – Panafrican to Hercynian deformations in the Mauretanides and tectonic significance of gravity anomalies. Tectonophysics, 109: 41–59

Rowley, D.B., Raymond, A., Parrish, J.T., Lottes, A.L., Scotese, C.R. and Ziegler, A.M., 1985 – Carboniferous paleogeographic, phytogeographic and paleoclimate reconstructions. Int. J. of Coal Geology, 5: 7–42

Rudkevich, M.Y., 1976 – The history and dynamics of the development of the West Siberian platform. Tectonophysics, 36 (1–3): 275–287

Rust, B.R., 1981 – Alluvial Deposits and Tectonic style: Devonian and Carboniferous successions in Eastern Canada. In: A.D. Miall (ed.). Sedimentation and Tectonics in Alluvial Basins. Geol. Ass. Can. Special Paper 23: 49–76

Ruzencev, S.V. and Samygin, S.G., 1979 – Die tektonische Entwicklung des Südurals im unteren und mittleren Paläozoikum. Z. geol. Wiss., Berlin, 7 (10): 1123–1186

Samygin, S.G. and Khain, Ye.V., 1985 – The South Urals and North Caucasus in the Paleozoic, an experiment in correlation. Geotectonics 19, (2): 113–126

Sandberg, C.A. and Maple, W.J., 1967 – Devonian of the Northern Rocky Mountains and plains. In: D.H. Oswald (ed.). International Symposium on the Devonian System Calgary, Alberta. Soc. Petrol. Geol., vol. 1: 843–879

Sanford, B.V., 1987 – Paleozoic geology of the Hudson Platform. In: C. Beaumont and T. Tankard (eds.). Basins of Eastern Canada, and worldwide analogues. Can. Soc. Petrol. Geol. Mem. 12: 483–505

Sanford, B.V. and Norris, A.W., 1973 – The Hudson Platform. In: McCrossan (ed.). Future Petroleum Provinces of Canada. Can. Soc. Petrol. Geol. Mem., 1: 387–409

Scheckler, S.E., 1986 – Old Red continent facies in the Late Devonian and Early Carboniferous of Appalachian North America. Ann. Soc. géol. Belgique, 109: 223–236

Schenk, P.E., 1978 – Synthetis of the Canadian Appalachians. In: IGCP project 27 Caledonian-Appalachian Orogen of the North Atlantic Region. Geol. Surv. Canada, Paper 78–13: 111–136

Schenk, P.E., 1982 – The Meguma Terrane of Nova Scotia Canada – an aid in Trans-Atlantic correlation. In : P.E. Schenk (ed.). Regional Trends in the Geology of the Appalachian-Caledonian-Hercynian-Mauretanide Orogen, D. Reidel Publ. Co.: 121–130

Schmidt, K. and Franke, D., 1975 – Stand und Probleme der Karbonforschung in der DDR. Z. geol. Wiss., Berlin, 3: 819–849

Schopf, T.J.M., 1974 – Permo-Triassic extinctions: relation to sea floor spreading. J. Geol., 82: 129–143

Scotese, C.R., Bambach, K.R., Barton C., van der Voo, R. and Ziegler, A.M., 1979 – Paleozoic basemaps. J. Geology, 87: 217–277

Scotese, C.R., van der Voo, R., Johnson, R.E. and Giles, P.S., 1984 – Paleomagnetic results from the Carboniferous of Nova Scotia. In: R. van der Voo, C.R. Scotese and N. Bonhommet (eds.). Plate reconstruction from Paleozoic Paleomagnetism. Geodynamics Series, vol. 12, Am. Geophys. Union; Soc. Am., Washington, 63–81

Scotese, C.R., Van der Voo, R. and Barrett, S.F., 1985 – Silurian and Devonian base maps. Phil. Trans. R. Soc. Lond., B 309: 57–77

Secor, D.T. jr., Snoke, A.W. and Dallmeyer, R.D., 1986 – Character of the Alleghanian orogeny in the southern Appalachians: Part III, regional tectonic relations. Geol. Soc. Am. Bull., 97: 1345–1353

Séranne, M. and Séguret, M., 1987 – The Devonian basins of Western Norway: tectonics and kinematics of an extending crust. In: M.P. Coward, J.F. Dewey and P.L. Hancock (eds.). Continental Extensional Tectonics. Geol. Soc. Lond. Spec. Publ. 28, Blackwell Sci. Publ., 537–548

Shurbet, D.H. and Cebull, S.E., 1987 – Tectonic interpretation of the westernmost part of the Ouachita-Marathon (Hercynian) orogenic belt, west Texas, Mexico. Geology, 15: 458–461

Skehan, J.W., 1983 – Geological profiles through the Avalonian terrain of southeastern Massachussetts, Rode Island and eastern Connecticut, USA. In: N. Rast and F.M. Delany (eds.). Profiles of orogenic belts. Geodynam. Series 10, Am. Geophys. Union; Geol. Soc. Am., Washington, 275–300

Skehan, J.W. and Rast, N., 1983 – Relationship between Precambrian and Lower Paleozoic rocks of southeastern New England and other North Atlantic Avalonian terraines. In: P.E. Schenk (ed.). Regional trends in the geology of the Appalachian-Caledonian-Hercynian--Mauretanide orogen. D. Reidel Publ. Co., 131–162

Skipp, B., 1979 – Great Basin Region. In: L.C. Craig and C. Waite-Connor (eds.). Paleotectonic Investigations of the Mississippian system in the United States, Part I: Introduction and regional analyses of the Mississippian system. Geol. Survey Professional Paper 1010: 271–328

Slezinger, A.E. and Jansin, A.L., 1979 – Die struktuelle Position der jungpaläoischen Depressionfazies im südlichen Vorural. Z. geol. Wiss. Berlin, 7 (10): 1187–1195

Sloss, L.L., 1963 – Sequences in the cratonic interior of North America. Geol. Soc. Am. Bull., 74: 93–113

Sloss, L.L., 1984 – Comparative anatomy of cratonic unconformities. In: J.S. Schlee (ed.). Interregional unconformities and hydrocarbon accumulation. Am. Assoc. Petrol. Geol. Mem. 36: 1–7

Smith, G.P. and Stearn, C.W., 1982 – The Devonian-Carbonate-Clastic sequence on Southwest Ellesmere Island, Arctic Canada. In: A.E. Embry and H.R. Balkwill (eds.). Arctic Geology and Geophysics, Can. Soc. Petrol. Geol. Mem. 8: 147–154

Smith, G.P. and Okulitch, A.V., 1987 – The Inglefield uplift: a Devonian tectonic element, Ellesmere Island Arctic Canada. Bull. Can. Soc. Petrol. Geol., 35 (1): 75–78

Sobolev, I.D., 1982 – Chaîne plisse de l'Oural. In: A.V. Peive, K.E. Khain, M.V. Mouratov and F. Delany (eds.). Tectonic of Europe and adjacent areas. Nauka Publishing House, Moscow, 185–213

Soper, N.J. and Hutton, D.H.W., 1984 – Late Caledonian sinistral displacements in Britain: Implications for a threeplate collision model. Tectonics, 3 (1): 781–794

Soper, N.J., Webb, B.C. and Woodcock, N.H., 1987 – Late Caledonian (Acadian) transpression in north-west England: timing, geometry and geotectonic significance. Proceed. Yorkshire Geol. Soc. 46 (3): 175–192

Southerland, P.K., 1988 – Late Mississippian and Pennsylvanian depositional history of the Arkoma basin, Oklahoma and Arkansas. Geol. Soc. Am. Bull. 100: 1787–1802

Spariosu D.J. and Kent D.V., 1983 – Paleomagnetism of the Lower Devonian Traveler Felsite and the Acadian orogeny in the New England Appalachians. Geol. Soc. Am. Bull., 94: 1319–1328

Spariosu, D.J., Kent, D.V. and Keppie, J.D., 1984 – Late Paleozoic motions of the Meguma terrane, Novan Scotia: new paleomagnetic evidence. In: R. van der Voo, C.R. Scotese and N. Bonhommet (eds.). Plate reconstruction from Paleozoic paleomagnetism. Geodynamic Series vol. 12, Am. Geophys. Union; Geol. Soc. Am., Washington, 82–97

Speed, R.C. and Sleep, N.H., 1982 – Antler orogeny and foreland basin: a model. Geol. Soc. Am. Bull. 93: 815–828

Steel, R.J., 1976 – Devonian basins of West Norway: sedimentary response to tectonism and varying tectonic context. Tectonophysics 36: 207–224

Steel, R.J. and Gloppen, T.G., 1980 – Late Caledonian (Devonian) basin formation, Western Norway: signs of strike slip tectonics during infilling. Spec. Publ. int. Ass. Sediment. 4: 79–103

Steel, R.J. and Worsley, D., 1984 – Svalbard's post-Caledonian strata – an atlas of sedimentational patterns and palaeographic evolution. In: A.M. Spencers et al. (eds.). Petroleum geology of the north European margin. Norwegian Petrol. Soc., Graham & Trotman Ltd., 109–136

Steel, R., Siedlecka, A. and Roberts, D., 1985 – The Old Red Sandstone basins of Norway and their deformation: a review. In: D.G. Gee and B.A. Sturt (eds.). The Caledonide Orogen-Scandinavia and Related Areas. John Wiley & Sons. Part 1: 293–317

Stemmerik, L. and Sørensen, M., 1980 – Upper Permian dykes in southern Scoresby land, East Greenland. Grønland Geol. Unders. Rapp. 100: 108

Stephenson, R.A., Embry, A.F., Nakiboglu, S.M. and Hastoaglu, M.A., 1987 Rift-initiated Permian to Early Cretaceous subsidence of the Sverdrup basin. In: C. Beaumont and A.J. Tankard (eds.). Sedimentary basins and basin forming mechanisms. Can. Soc. Petrol. Geol. Mem. 12: 213–231

Stevens, C.H. and Stone, P., 1988 – Early Permian thrust faults in east-central California. Geol. Soc.Am. Bull. 100: 552–562

St. John, B., 1986 – Antarctica-geology and hydrocarbon potential. In: M.T. Halbouty (ed.). Future petroleum provinces of the world. Am. Assoc. Petrol. Geol. Mem. 40: 55–100

Stockmal, G.S., Beaumont, C. and Boutilier, R., 1986 – Geodynamic models of convergent margin tectonics: transition from rifted margin to overthrust belt and consequences for foreland-basin development. Am. Assoc. Petrol. Geol. Bull. 70: 181–190

Sturt, B.A., Pringle, I.R. and Ramsey, D.M., 1978 – The Finmarkian phase of the Caledonian Orogeny. J. Geol. Soc. Lond., 135: 597–610

Surlyk F. and Hurst J.M., 1984 – The evolution of the early Paleozoic deep water basin of North Greenland. Geol. Soc. Am. Bull. 95: 131–154

Surlyk, F., Hurst, J.M., Piaseki, S., Rolle, F., Scholle, P.A., Stemmerik, L. and Thomsen, E., 1986 – The Permian of the western margin of the Greenland Sea a future exploration target. In: M.T. Halbouty (ed.). Future petroleum provinces of the world. Am. Assoc. Petrol. Geol. Mem. 40: 629–659

Svoboda J., 1966 – The Barrandian Basin. In: J. Svoboda (ed.). Regional Geology of Czechoslovakia, Part I, The Bohemian Massif. Geol. Survey Czechoslovakia, Publ.: 281–341

Sweeney, J.F., 1977 – Subsidence of the Sverdrup basin, Canadian Arctic. Geol. Soc. Am. Bull. 88: 41–48

Tailleur, I.L., Brosgé, W.P. and Reiser, H.N., 1967 – Palinspastic analysis of Devonian rocks in N.W. Alaska. In: D.H. Oswald (ed.). International Symp. on the Devonian System, Alberta. Soc. Petrol. Geol., vol. 2: 1345–1361

Tankard, A.J., 1986 – On the depositional response to thrusting and lithospheric flexture: examples from the Appalachian and Rocky Mountain basins. Spec. Publ. Int. Ass. Sediment, 8: 369–392

Tate, M.P. and Dobson, M.R., 1989 – Pre-Mesozoic geology of the western and north-western Irish continental shelf. J. Geol. Soc. Lond. 146: 229–240

Tate, M.P. and Dobson, M.R., 1989 – Late Permian to early Mesozoic rifting and sedimentation offshore NW Ireland. Mar. Petrol. Geol. 6 (1): 49–59

Tauvers, P.R. and Muehlberger, 1987 – Is the Brunswick magnetic anomaly really the Alleghenian suture? Tectonics 6: 331–342

Telford, P.G., 1988 – Devonian stratigraphy of the Moose River Basin, James Bay Lowlands, Ontario, Canada. In: N.J. McMillan, A.E. Embry and D.J. Glass (eds.). Devonian of the World. Can. Soc. Petrol. Geol. Mem. 14, Vol. 1: 123–132

Thiele, O., 1977 – Zur Tektonik des Waldviertels in Niederösterreich (südliche Böhmische Masse). Nova Acta Leopoldina NF 244: 67–82

Thorpe, J.E., 1988 – The Devonian of the Hudson Bay Basin. In: N.J. McMillan, A.E. Embry and D.J. Glass (eds.). Devonian of the World. Can. Soc. Petrol. Geol. Mem. 14, Vol.1: 133–154

Tollman, A., 1984 – Entstehung und früher Werdegang der Tethys mit besonderer Berücksichtigung des Mediterranen Raumes. Mitt. Österr. geol. Ges. 77: 93–113

Tomczyk, H., 1970 – The Silurian. In: Geology of Poland, vol. 1, Stratigraphy, part 1, Precambrian and Palaeozoic. Publ. House Wydawnictwa, Geologiczne, Warsawa, 237–320

Tomczik, H. and Tomczikova, E., 1981 – Korelacja Biostratygraficzna Silura W.

Europie. Przeglad geologiczny XXIX 6 (338): 283–291

Torsvik, T.H., Sturt, B.A., Ramsey, D.M. and Velti, V., 1987 – The tectono-magnetic signature of the old red sandstone and pre-Devonian strata in the Håsteinen area, Western Norway, and implications for the later stages of the Caledonian Orogeny. Tectonics, 6 (3): 305–322

Trettin H.P., 1987 – Pearya: a composite terrane with Caledonian affinities in northern Ellesmere Island. Can. J. Earth Sci. 24: 224–245

Trettin, H.P. and Balkwill, H.R., 1979- Contributions to the tectonic history of the Inuitian Province, Arctic Canada. Can. J. Earth Sci. 16 (3): 748–769

Trettin, H.P., Frisch, T.O., Sobczak, C.W., Weber, J.R., Law, L.K., Delaurier, I., Niblett, F.R. and Whitham, K., 1972 – The Inuitian Province. In: R.A. Price and R.J.W. Douglas (eds.). Variations in tectonic styles in Canada. Geol. Assoc. Canada Spec. Paper 11: 83–179

Trettin, H.P., Parrish, R. and Loveridge, W.D., 1987 – U-Pb age determination on Proterozoic to Devonian rocks from northern Ellesmere Island, Arctic Canada. Can. J. Earth Sci. 24: 246–256

Trueman, A.E., 1946 – Stratigraphical problems in the coal measures of Europe and North America. Q. Jl. Geol. Soc. CII: 1-XCIII

Tull, J.F. and Telle, W.R., 1988 – Tectonic setting of the Silurian (?)-Devonian Talladega Belt successor basin, southern Appalachians. In: N.J. McMillan, A.F. Embry and D.J. Glass (eds.). Devonian of the World. Can. Soc. Petrol. Geol. Mem. 14, Vol. 2: 209–224

Ulmishek, G., 1982 – Petroleum geology and resource assessment of the Timan-Pechora basin, USSR, and the adjacent Barents-northern Kara Shelf. Argonne National Laboratory, US Department of Commerce, National Technical Information Service

DE83-004421, 197 p

Ulmishek, G., 1984 – The geology and petroleum resources of basins in the Asian Arctic and offshore East Greenland. Argonne National Laboratory, 9700 South Cass Avenue Argonne, Illinois 60439, 104 p

Ulmishek, G., 1985 – Geology and petroleum resources of the Barents – North Kara shelf in the light of new geological data. Argonne National Laboratory, ANL/ES–148, 89 p

Ustritsky, V.I., 1973 – Permian climate. In: A. Logan and L.V. Hills (eds.). The Permian and Triassic systems and their mutual boundary. Can. Soc. Petrol. Geol. Mem. 2: 733–744

Uyeda, S., 1981 – Subduction zones and back-arc basins: a review. Geol. Rundsch., 70 (2): 552–569

Vai, G.B. and Cocozza, T., 1986 – Tentative schematic zonation of the Hercynian chain in Italy. Bull. Soc. géol. France (8) II (1): 95–114

Vail, P.R., Mitchum, R.M. jr., Todd, R.G., Widmier, J.M., Thompson, S., Sangree, J.B., Bubb, J.N. and Hatledid, W.G., 1977 – Seismic stratigraphy and global changes of sea level. In: C.E. Payton (ed.). Seismic stratigraphy application to hydrocarbon exploration. Am. Assoc. Petrol. Geol. Mem. 26: 49–212

Vauchez, A., Kessler, S.F., Lchorche, J.P. and Villeneuve, M., 1987 – Southward extrusion tectonics during the Carboniferous Africa-North America collision. Tectonophysics 142: 317–322

Vinogradov, A.P. (ed.), 1969 – Atlas of the lithological-palaeogeographical maps of the USSR, vol. 2, Devonian, Carboniferous and Permian. Akad. Sci. USSR, Moscow

Vischer, A., 1943 – Die postdevonische Tektonik von Ostgrønland zwischen 740° und 750°N Br. Kuhn T, Wollaston Forland, Clavering T und angrenzender Gebiete. Meddr. Grønland 133: 1–195

Volker, F. and Altherr, R., 1987 – Lower Carboniferous calcalkaline volcanics in the northern Vosges: evidence for a destructive continental margin. Terra Cognita, 7 (2–3): 174–175 (abstract)

Voo R. van der, 1988 – Paleozoic Paleogeography of North America, Gondwana and intervening displaced Terranes: comparison of paleomagnetism with paleoclimatology and biogeographical Patterns. Geol. Soc. Am. Bull. 100: 311–324

Walker, J.D., 1988 – Permian and Triassic rocks of the Mojave desert and their implication for timing and mechanisms of continental truncation. Tectonics 7 (3): 685–709

Walliser, O.H. and Michels, D., 1983 – Der Ursprung des Rheinischen Schelfes im Devon. N-Jb. Geol. Paläont. Abh. 166 (1): 3–18

Walper, J.L., 1981 – Geological evolution of the Gulf of Mexico-Caribbean region. In: J.W. Kerr, A.J. Fergusson and L.C. Machan (eds.). Geology of the North Atlantic borderlands. Can. Soc. Petrol. Geol. Mem. 7: 503–525

Watson, J., 1985 – Northern Scotland as an Atlantic-North Sea divide. J. Geol. Soc. Lond., 142 (2): 221–243

Watts, A.B., Karner, G.D. and Steckler, M.S., 1982 – Lithospheric flexure and the evolution of sedimentary basins. In: Sir Peter Kent, M.H.P. Bott, D.P. McKenzie and C.A. Williams (eds.). The evolution of sedimentary basins. Phil. Trans., Roy. Soc. London, Ser. A, vol. 305: 249–282

Watznauer, A., Tröger, K.A. and Möbus, G., 1976 – Gleichheiten und Underschiede im Bau der Saxothuringischen Zone westlich und östlich des Elbelineaments. In: A. Watznauer (ed.). Franz Kossmat Symposium. Nova Acta Leopldina N.F. 224: 93–110

Weber, K., 1984 – Variscan events: early Palaeozoic continental rift metamorphism and Late Palaeozoic crustal shortening. In: D.H.W. Hutton and D.J. Sanderson (eds.).

Variscan tectonics of the North Atlantic region. Geol. Soc. Lond. Spec. Publ. 14: 3–22

Wendt, J., 1985 – Disintegration of the continental margin of northwestern Gondwana: Late Devonian of the eastern Anti-Atlas (Morocco). Geology, 13: 815–818

Wendt J. 1988 – Facies pattern and paleogeography of the Middle and Late Devonian in the eastern Anti-Atlas (Morocco). In: N.J. McMillan, A.F. Embry and D.J. Glass (eds.). Devonian of the World. Can. Soc. Petrol. Geol. Mem. 14, Vol. 1: 467–480

Wickham, J., Roeder, D. and Briggs, G., 1976 – Plate tectonics models for the Ouachita foldbelt. Geology 4 (3): 173–176

Williams, H. and Hatcher, R.D. jr., 1983 – Appalachian suspect terranes. Geol. Soc. Am. Mem., 158: 33–53

Wilson, C.W. jr. and Stearn, R.G., 1963 – Quantitative analysis of Ordovician and younger structural development of Nashville Dome, Tennessee. Am. Assoc. Petrol. Geol. Bull., 47: 823–832

Wilson, J.T., 1966 – Did the Atlantic close and then reopen? Nature, London, 211: 676–681

Wintsch, R.P. and Sutter, J.F., 1986 – A tectonic model for the Late Paleozoic of southeastern New England. J. Geol. 94: 459–472

Witkowski, A., 1979 – Atlas Geologiczno – structuralny Poludniowej Czesci Morze Baltyckiego. Wydawnictwa Geologiczne, Warsawa

Woodcock, N.H., Awan, M.A., Johnson, T.E., Mackie, A.H. and Smith, R.D.A., 1988 – Acadian tectonics of Wales during Avalon/Laurentia convergence. Tectonics 7 (3): 483–495

Woodrow, D.L., Fletcher, F.W. and Ahrnsbrak, W.F., 1973 – Paleogeography and Paleoclimate at the Deposition Sites of the Devonian Catskill and Old Red Facies. Geol. Soc. Am. Bull. 84: 3051–3064

Wuellner, D.E., Lehtonen, L.R. and James,

W.C., 1986 – Sedimentary-tectonic development of the Marathon and Val Verde basins, West Texas, USA: a Permo-Carboniferous migrating foredeep. Spec. Publs. Int. Ass. Sediment 8: 347–368

Wynne, P.J., Irving, E. and Osadete, K., 1983 – Paleomagnetism of the Esayoo Formation (Permian) of northern Ellesmere Island: possible clue to the solution of the Nares Strati dilemma. Tectonophysics 100: 241–256

Young, G.C., 1987 – Devonian palaeonotological data and the Armorican problem. Palaeogeogr., Palaeoclim., Palaeocol., 60: 283–304

Zeck, H.P., Andriessen, P.A.M., Hansen, K., Jensen, P.K. and Rasmussen, B.L., 1988 – Paleozoic paleo-cover of the southern part of the Fennoscandian Shield-fisson track constraints. Tectonophysics 149 (1–2): 61–66

Ziegler, A.M., 1981 – Paleozoic paleogeography. In: M.W. McElhinny, D.A. Valencio (eds.). Paleoreconstruction of the continents. Geodynamic Series vol 2, Am. Geophys. Union and Geol. Soc. Am. 31–37

Ziegler, A.M., Scotese, C.R., McKerrow, W.S., Johnson, M.E. and Bambach, R.K., 1979 – Paleozoic paleogeography. Ann. Rev. Earth Planet Sci. 7: 473–502

Ziegler, P.A., 1969 – The development of sedimentary basins in western and Arctic Canada. Alberta Soc. Petrol. Geol.

Ziegler, P.A., 1982 – Geological atlas of Western and Central Europe. Elsevier, Sci. Publ. Co., Amsterdam, 130 p and 40 plates

Ziegler, P.A., 1984 – Caledonian and Hercynian crustal consolidation of Western and Central Europe – a working hypothesis. Geol. Mijnbouw, 63 (1): 93–108

Ziegler, P.A., 1986 – Geodynamic model for Palaeozoic crustal consolidation of Western and Central Europe. Tectonophysics, 126: 303–328

Ziegler, P.A., 1988 – Evolution of the Arctic-North Atlantic and Western Tethys. Am. Assoc. Petrol. Geol. Mem. 43: 198 p. and 30 plates

Ziegler, P.A., 1988 – Laurussia – the Old Red Continent. In: N.J. McMillan, A.F. Embry and D.J. Glass (eds.). Devonian of the World. Can. Soc. Petrol. Geol. Mem. 14, Vol. 1: 15–48

Ziegler, P.A. and van Hoorn, B., 1989 – Evolution of the North Sea rift system. In: A.J. Tankard and H. Balkwill (eds.). Extensional tectonics and stratigraphy of the North Atlantic. Am. Ass. Petrol. Geol. Mem. 43 (in press)

Zoback, M.L. and Zoback, M.D., 1980 – State of stress in the conterminous United States. J. Geophys. Res. 85: 6113–6156

Zonenshain, L.P., Korinevsky, V.G., Kuzmin, V.G. and Pechersky, D.M., 1984 – Plate tectonic model of the south Urals development. Tectonophysics, 109: 95–135

Zonenshain, L.P., Kuzmin, M.I. and Natapov, L.M., 1987a – New outlook on geological history of the Soviet Union. Priroda (Nature), General Academy of Science USSR, 1987 (2): 32–40 (in Russian)

Zonenshain, L.P., Kuzmin, M.I. and Konov, M.V., 1987b – Absolute reconstruction of continent configuration during the Palaeozoic and Early Mesozoic. Geotektonika, 1987 (2): 16–27 (in Russian)

Zwart, H.J. and Dornsiepen, V.F., 1978 – The tectonic framework of central and western Europe. Geologie Mijnb. 57: 627–654

Plate 1

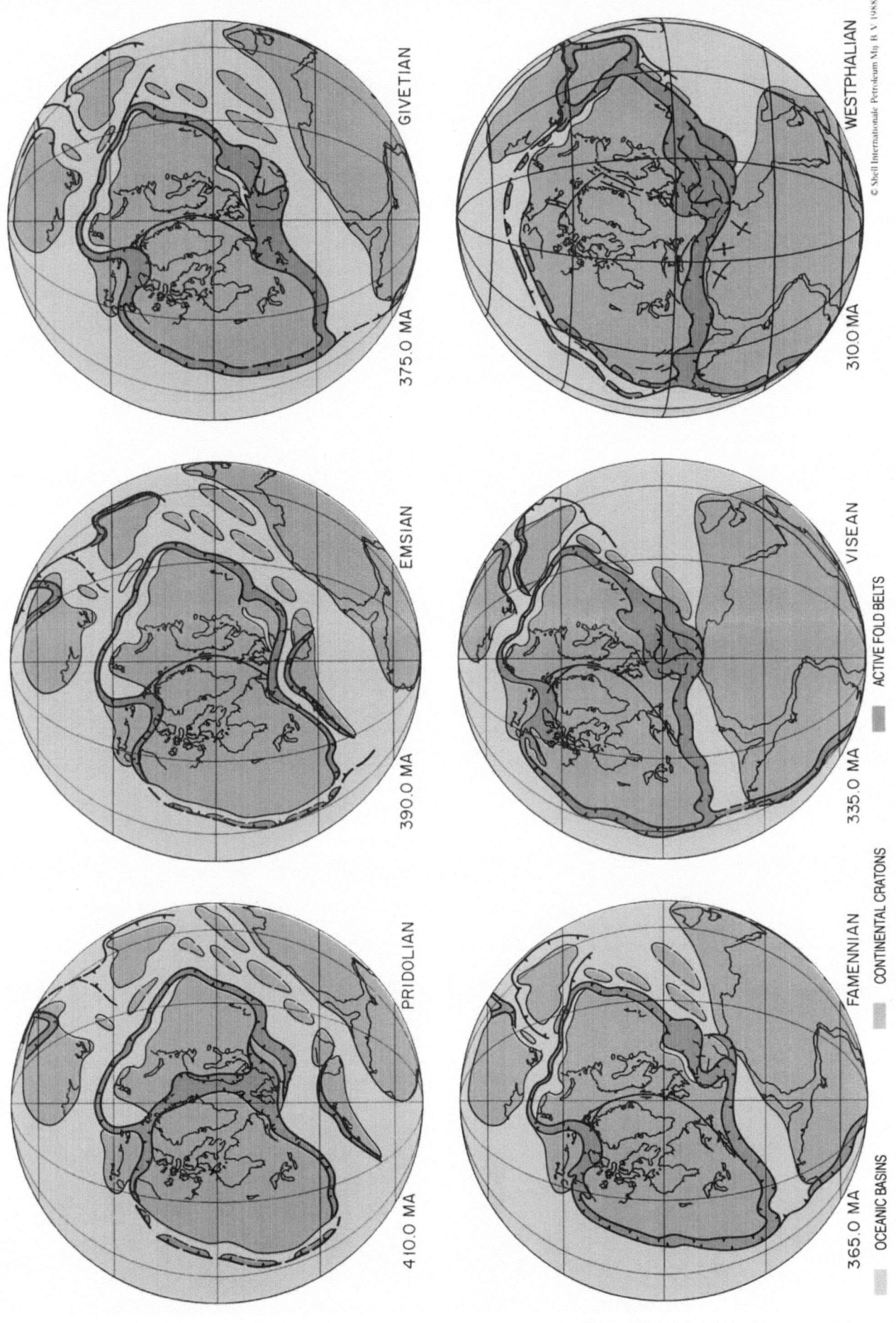

GIVETIAN   375.0 MA

EMSIAN   390.0 MA

PRIDOLIAN   410.0 MA

WESTPHALIAN   310.0 MA

© Shell Internationale Petroleum Mij B.V 1988

VISEAN   335.0 MA

FAMENNIAN   365.0 MA

ACTIVE FOLD BELTS

CONTINENTAL CRATONS

OCEANIC BASINS

Plate 2

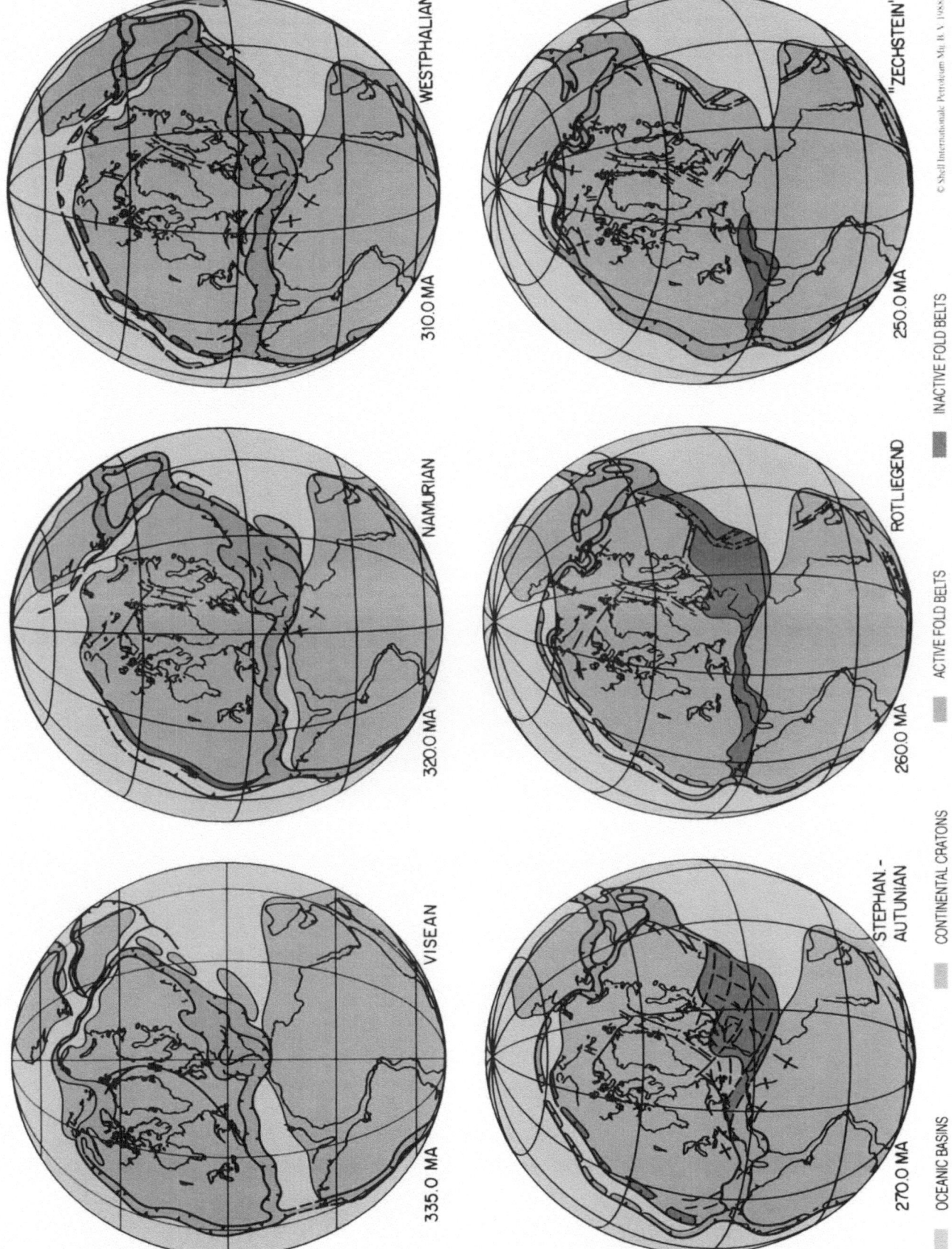

WESTPHALIAN
310.0 MA

NAMURIAN
320.0 MA

VISEAN
335.0 MA

"ZECHSTEIN"
250.0 MA

ROTLIEGEND
260.0 MA

STEPHAN -
AUTUNIAN
270.0 MA

© Shell Internationale Petroleum Mij. B.V. 1988

INACTIVE FOLD BELTS

ACTIVE FOLD BELTS

CONTINENTAL CRATONS

OCEANIC BASINS

Plate 3

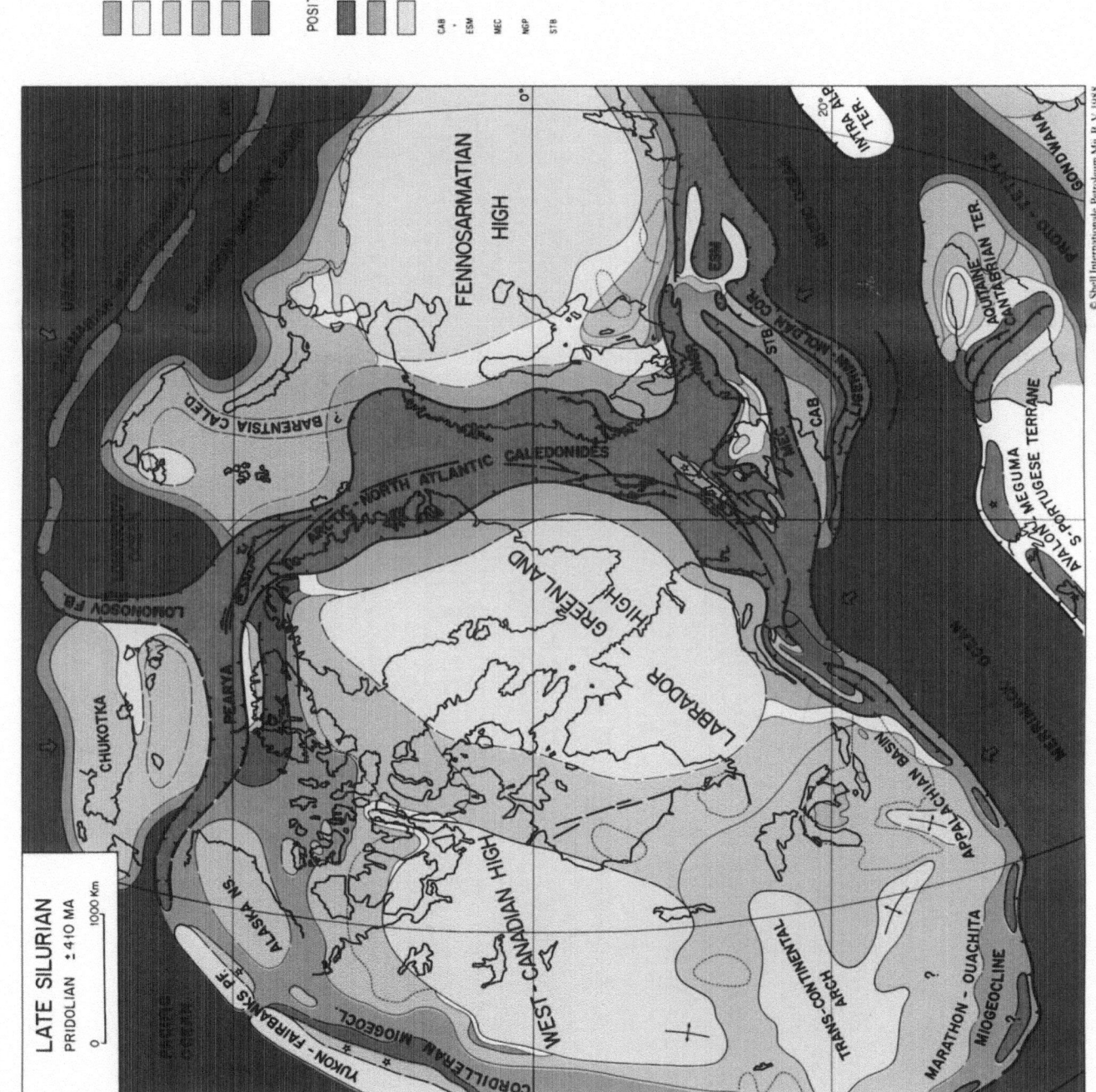

DEPOSITIONAL ENVIRONMENT AND PRINCIPAL LITHOLOGY

MAINLY CONTINENTAL CLASTICS
DELTAIC-SHALLOW MARINE, MAINLY SANDS
SHALLOW MARINE, MAINLY SHALES
SHALLOW MARINE, CARBONATES AND CLASTICS
SHALLOW MARINE, MAINLY CARBONATES
EVAPORITES AND CLASTICS

MAINLY EVAPORITES
EVAPORITES, CLASTICS AND CARBONATES
EVAPORITES AND CARBONATES
DEEPER MARINE CLASTICS AND/OR CARBONATES
DEEPER MARINE, MAINLY SANDS (FLYSCH)
BASINS FLOORED BY OCEANIC CRUST

POSITIVE AREAS

ACTIVE FOLD BELTS
INACTIVE FOLD BELTS
ANOROGENIC, CRATONIC

CONVERGENCE DIRECTION
ANOROGENIC VOLCANISM
OROGENIC VOLCANISM

CAB   CENTRAL ARMORICAN BASIN
ESM   EAST SILESIAN MASSIF
MEC   MID-EUROPEAN CALEDONIDES
NGP   NORTH GERMAN POLISH CALEDONIDES
STB   SAXOTHURINGIAN BASIN

LATE SILURIAN
PRIDOLIAN ± 410 MA

0   1000 Km

© Shell Internationale Petroleum Mij. B.V. 1988

Plate 4

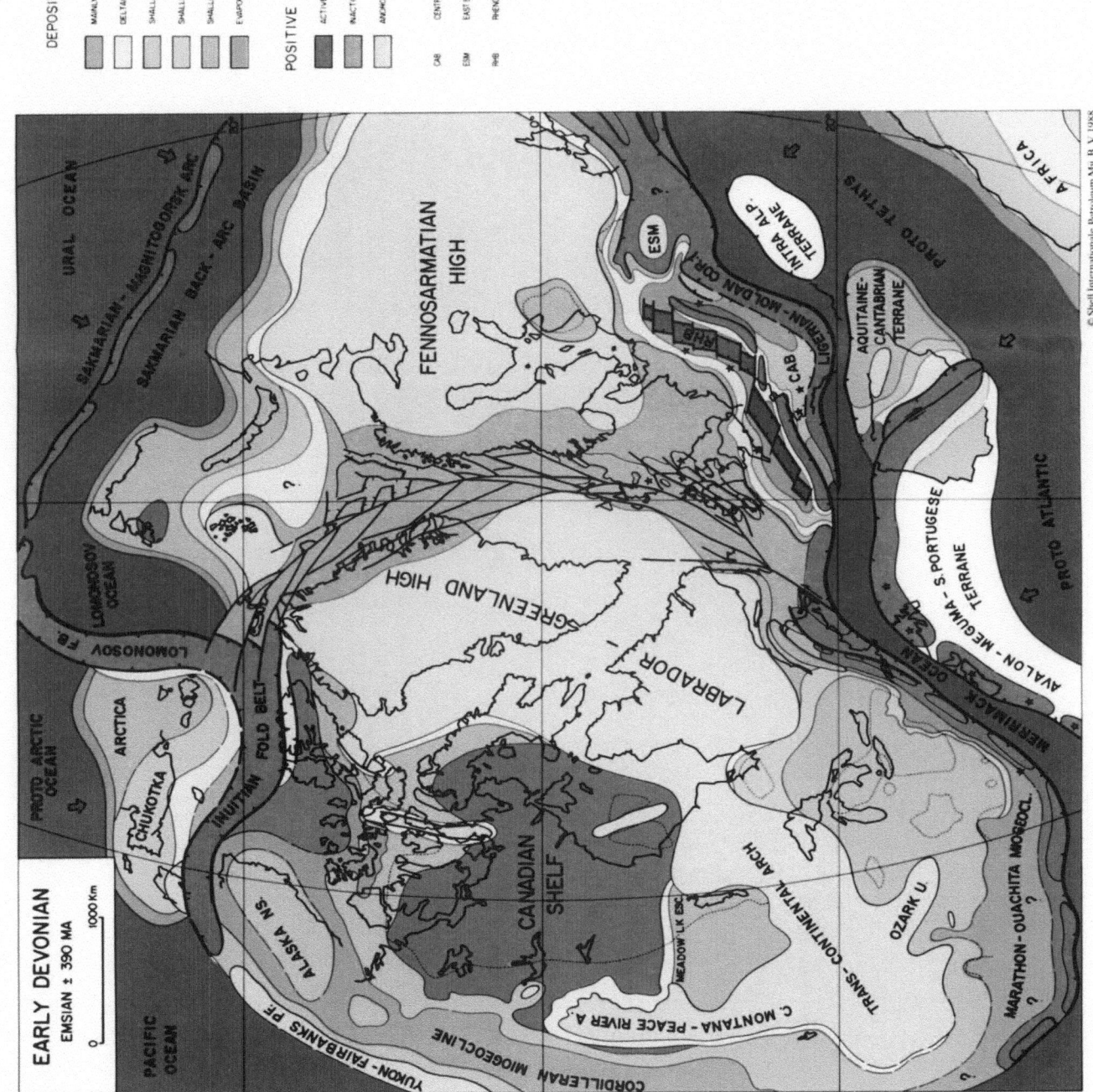

DEPOSITIONAL ENVIRONMENT AND PRINCIPAL LITHOLOGY

MAINLY CONTINENTAL CLASTICS

DELTAIC-SHALLOW MARINE, MAINLY SANDS

SHALLOW MARINE, MAINLY SHALES

SHALLOW MARINE, CARBONATES AND CLASTICS

SHALLOW MARINE, MAINLY CARBONATES

EVAPORITES AND CLASTICS

MAINLY EVAPORITES

EVAPORITES, CLASTICS AND CARBONATES

EVAPORITES AND CARBONATES

DEEPER MARINE CLASTICS AND/OR CARBONATES

DEEPER MARINE, MAINLY SANDS (FLYSCH)

BASINS FLOORED BY OCEANIC CRUST

CONVERGENCE DIRECTION

ANOROGENIC VOLCANISM

OROGENIC VOLCANISM

POSITIVE AREAS

ACTIVE FOLD BELTS

INACTIVE FOLD BELTS

ANOROGENIC, CRATONIC

CAB   CENTRAL ARMORICAN BASIN

ESM   EAST SILESIAN MASSIF

RHB   RHENOHERCYNIAN BASIN

EARLY DEVONIAN
EMSIAN ± 390 MA

0   1000 Km

© Shell Internationale Petroleum Mij B.V. 1988.

Plate 5

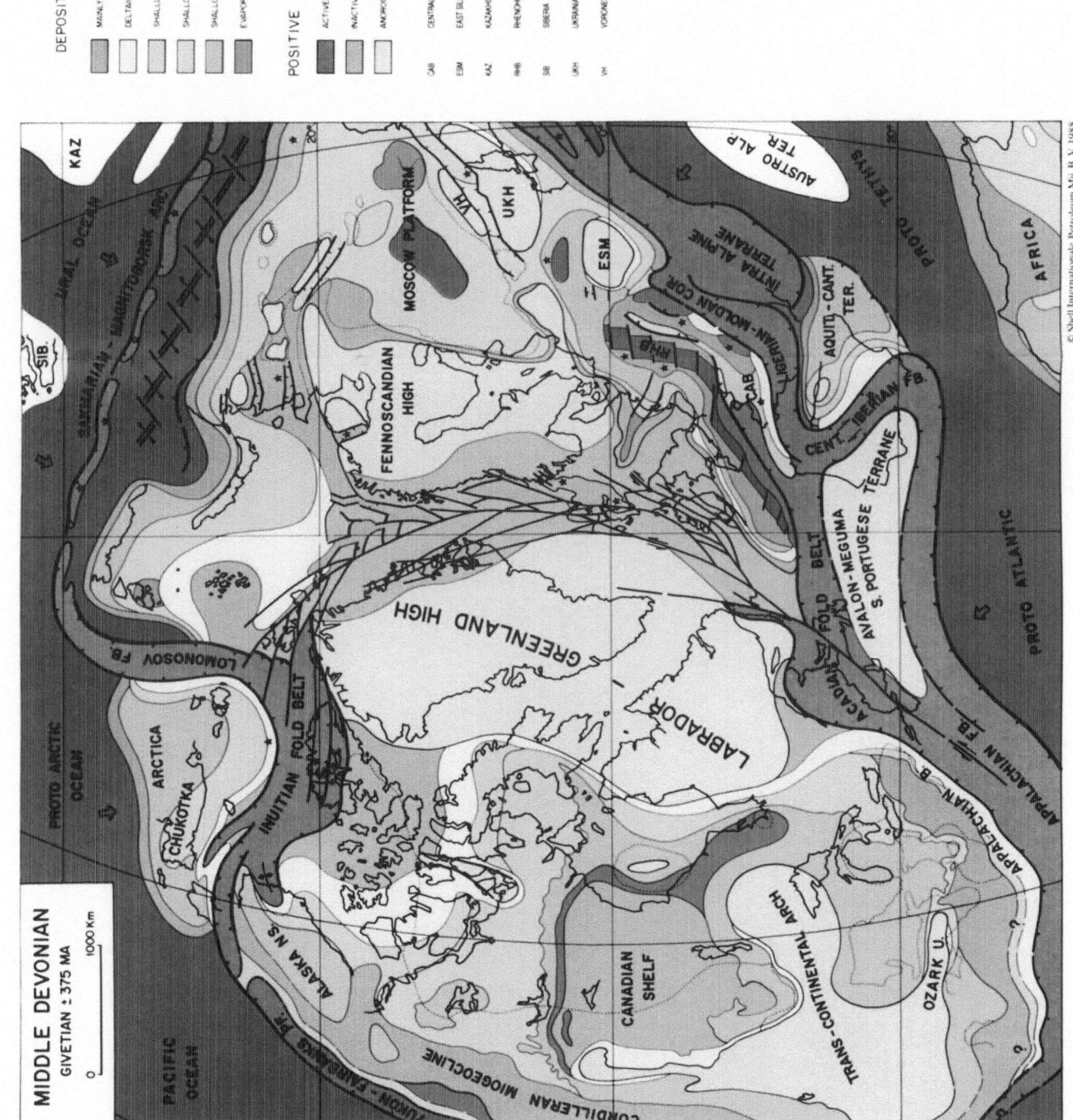

MIDDLE DEVONIAN
GIVETIAN ± 375 MA

0   1000 Km

DEPOSITIONAL ENVIRONMENT AND PRINCIPAL LITHOLOGY

MAINLY CONTINENTAL CLASTICS
DELTAIC SHALLOW MARINE, MAINLY SANDS
SHALLOW MARINE, MAINLY SHALES
SHALLOW MARINE, CARBONATES AND CLASTICS
SHALLOW MARINE, MAINLY CARBONATES
EVAPORITES AND CLASTICS

MAINLY EVAPORITES
EVAPORITES, CLASTICS AND CARBONATES
EVAPORITES AND CARBONATES
DEEPER MARINE CLASTICS AND/OR CARBONATES
DEEPER MARINE, MAINLY SANDS (FLYSCH)
BASINS FLOORED BY OCEANIC CRUST

CONVERGENCE DIRECTION
AMPHIBOGENIC VOLCANISM
OROGENIC VOLCANISM

POSITIVE AREAS

ACTIVE FOLD BELTS
INACTIVE FOLD BELTS
AMPHIBOGENIC, CRATONIC

CAB   CENTRAL ARMORICAN BASIN
ESM   EAST SILESIAN MASSIF
KAZ   KAZAKHSTAN
RHB   RHENOHERCYNIAN BASIN
SIB   SIBERIA
URH   URANIAN HIGH
VH   VORONESH HIGH

© Shell Internationale Petroleum Mij B V 1988

Plate 6

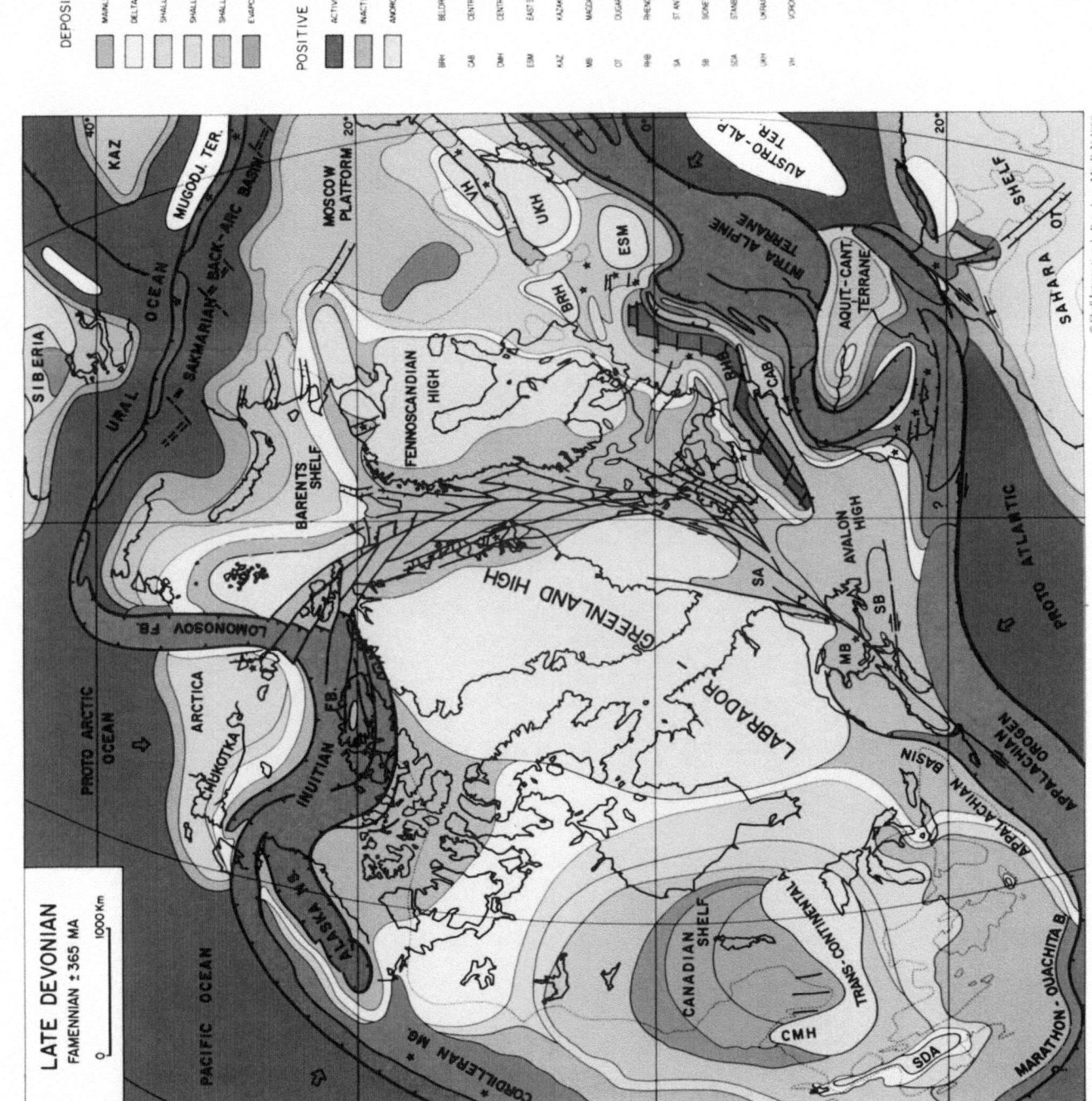

DEPOSITIONAL ENVIRONMENT AND PRINCIPAL LITHOLOGY

- MAINLY CONTINENTAL CLASTICS
- DELTAIC, SHALLOW MARINE, MAINLY SANDS
- SHALLOW MARINE, MAINLY SHALES
- SHALLOW MARINE, CARBONATES AND CLASTICS
- SHALLOW MARINE, MAINLY CARBONATES
- EVAPORITES AND CLASTICS

- MAINLY EVAPORITES
- EVAPORITES, CLASTICS AND CARBONATES
- EVAPORITES AND CARBONATES
- DEEPER MARINE CLASTICS AND/OR CARBONATES
- DEEPER MARINE, MAINLY SANDS (FLYSCH)
- BASINS FLOORED BY OCEANIC CRUST

↓ CONVERGENCE DIRECTION

✦ ANOROGENIC VOLCANISM

✩ OROGENIC VOLCANISM

POSITIVE AREAS

- ACTIVE FOLD BELTS
- INACTIVE FOLD BELTS
- ANOROGENIC, CRATONIC

BRH	BELORUSSIAN HIGH
CAB	CENTRAL ARMENIAN BASIN
CMH	CENTRAL MONTANA HIGH
ESM	EAST SILESIAN MASSIF
KAZ	KAZAKHSTAN
MB	MAGDALENE BASIN
OT	OUGARTA TROUGH
RHB	RHENOHERCYNIAN BASIN
SA	ST ANTHONY BASIN
SB	SIDNEY BASIN
SDA	STANBURY DEFIANCE ARCH
URH	UKRAINIAN HIGH
VH	VORONESH HIGH

LATE DEVONIAN
FAMENNIAN ± 365 MA

0   1000 Km

© Shell Internationale Petroleum Mij B V 1988

Plate 7

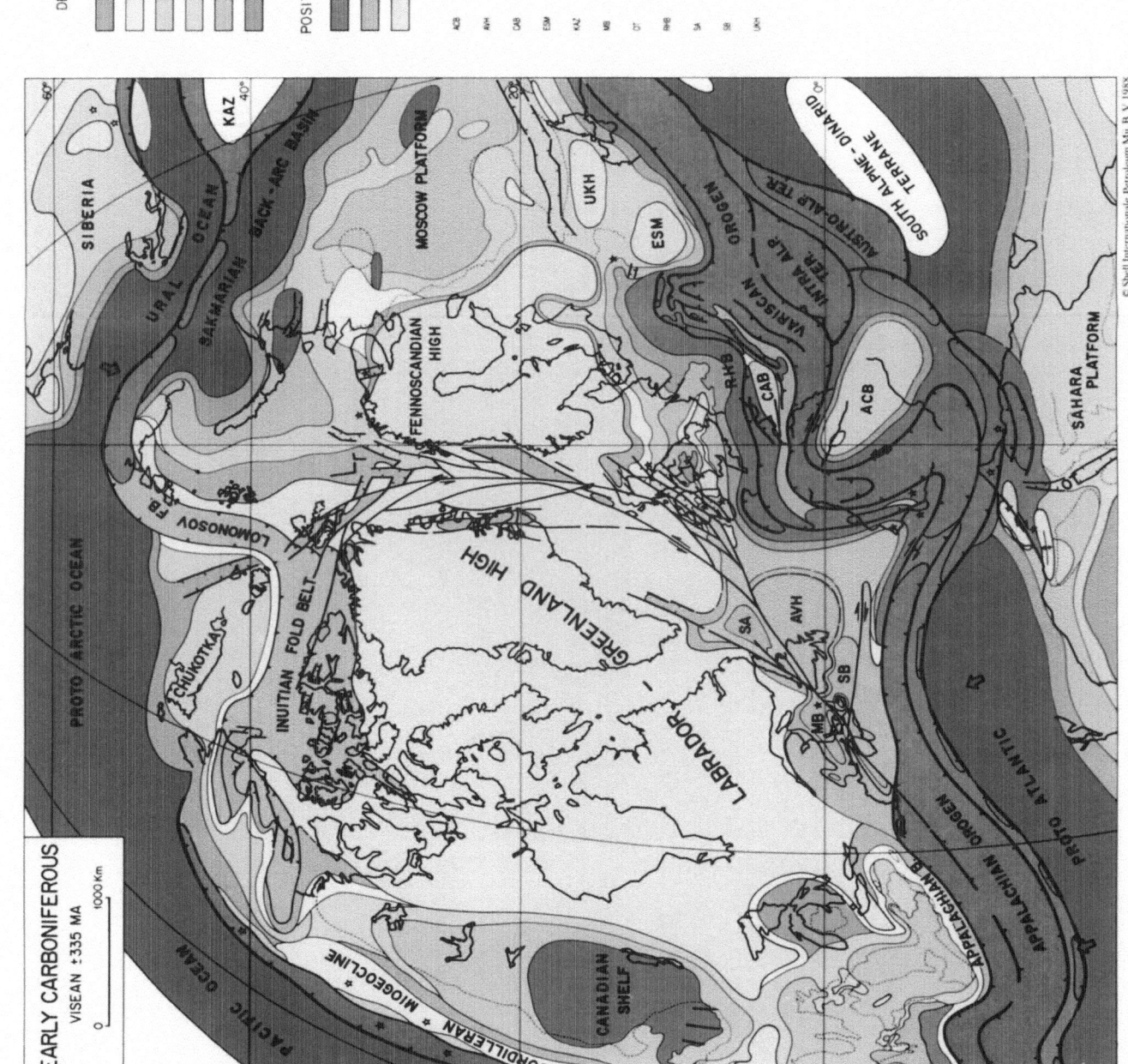

EARLY CARBONIFEROUS
VISEAN ±335 MA

0   1000 Km

DEPOSITIONAL ENVIRONMENT AND PRINCIPAL LITHOLOGY

MAINLY CONTINENTAL CLASTICS
DELTAIC: SHALLOW MARINE, MAINLY SANDS
SHALLOW MARINE, MAINLY SHALES
SHALLOW MARINE, CARBONATES AND CLASTICS
SHALLOW MARINE, MAINLY CARBONATES
EVAPORITES AND CLASTICS

MAINLY EVAPORITES
EVAPORITES, CLASTICS AND CARBONATES
EVAPORITES AND CARBONATES
DEEPER MARINE, CLASTICS AND/OR CARBONATES
DEEPER MARINE, MAINLY SANDS (FLYSCH)
BASINS FLOORED BY OCEANIC CRUST

CONVERGENCE DIRECTION
ANOROGENIC VOLCANISM
OROGENIC VOLCANISM

POSITIVE AREAS

ACTIVE FOLD BELTS
INACTIVE FOLD BELTS
ANOROGENIC, CRATONIC

ACB   AQUITAINE-CANTABRIA BASIN
AVH   AVALON HIGH
CAB   CENTRAL ARMORICAN BASIN
ESM   EAST SILESIAN MASSIF
KAZ   KAZAKHSTAN
MB   MAGDALENE BASIN
OT   OULGARTA TROUGH
RHB   PRE-HERCYNIAN BASIN
SA   ST. ANTHONY BASIN
SB   SIDNEY BASIN
UKH   UKRAINIAN HIGH

© Shell Internationale Petroleum Mij. B.V. 1988.

Plate 8

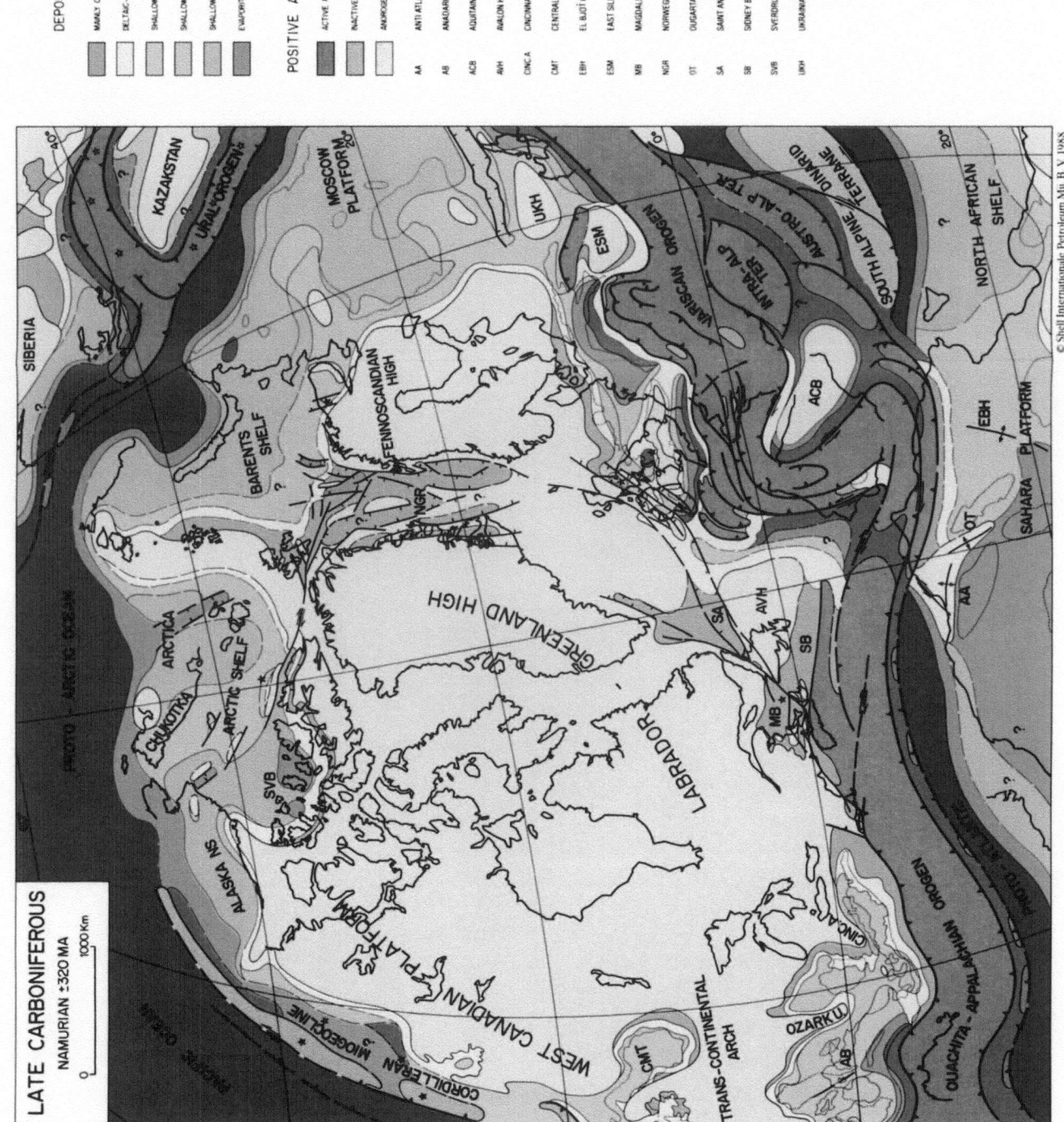

LATE CARBONIFEROUS
NAMURIAN ±320 MA

0   1000 Km

DEPOSITIONAL ENVIRONMENT AND PRINCIPAL LITHOLOGY

MAINLY CONTINENTAL CLASTICS
DELTAIC-SHALLOW MARINE, MAINLY SANDS
SHALLOW MARINE, MAINLY SHALES
SHALLOW MARINE, CARBONATES AND CLASTICS
SHALLOW MARINE, MAINLY CARBONATES
EVAPORITES AND CLASTICS

MAINLY EVAPORITES
EVAPORITES, CLASTICS AND CARBONATES
EVAPORITES AND CARBONATES
DEEPER MARINE, CLASTICS AND/OR CARBONATES
DEEPER MARINE, MAINLY SANDS (FLYSCH)
BASINS FLOORED BY OCEANIC CRUST

CONVERGENCE DIRECTION
ANOROGENIC VOLCANISM
OROGENIC VOLCANISM

POSITIVE AREAS

ACTIVE FOLD BELTS
INACTIVE FOLD BELTS
ANOROGENIC, CRATONIC

AA   ANTI ATLAS
AB   ANADARKO BASIN
ACB   AQUITAINE CANTABRIA BLOCK
AVH   AVALON HIGH
CINCA CINCINNATI ARCH
CMT   CENTRAL MONTANA TROUGH
EBH   EL BLÖT HIGH
ESM   EAST SILESIAN MASSIF
MB   MAGDALEN BASIN
NGR   NORWEGIAN GREENLAND SEA RIFT
OT   OUGARTA TROUGH
SA   SAINT ANTHONY BASIN
SB   SIDNEY BASIN
SVB   SVERDRUP BASIN
UKH   UKRAINIAN HIGH

Plate 9

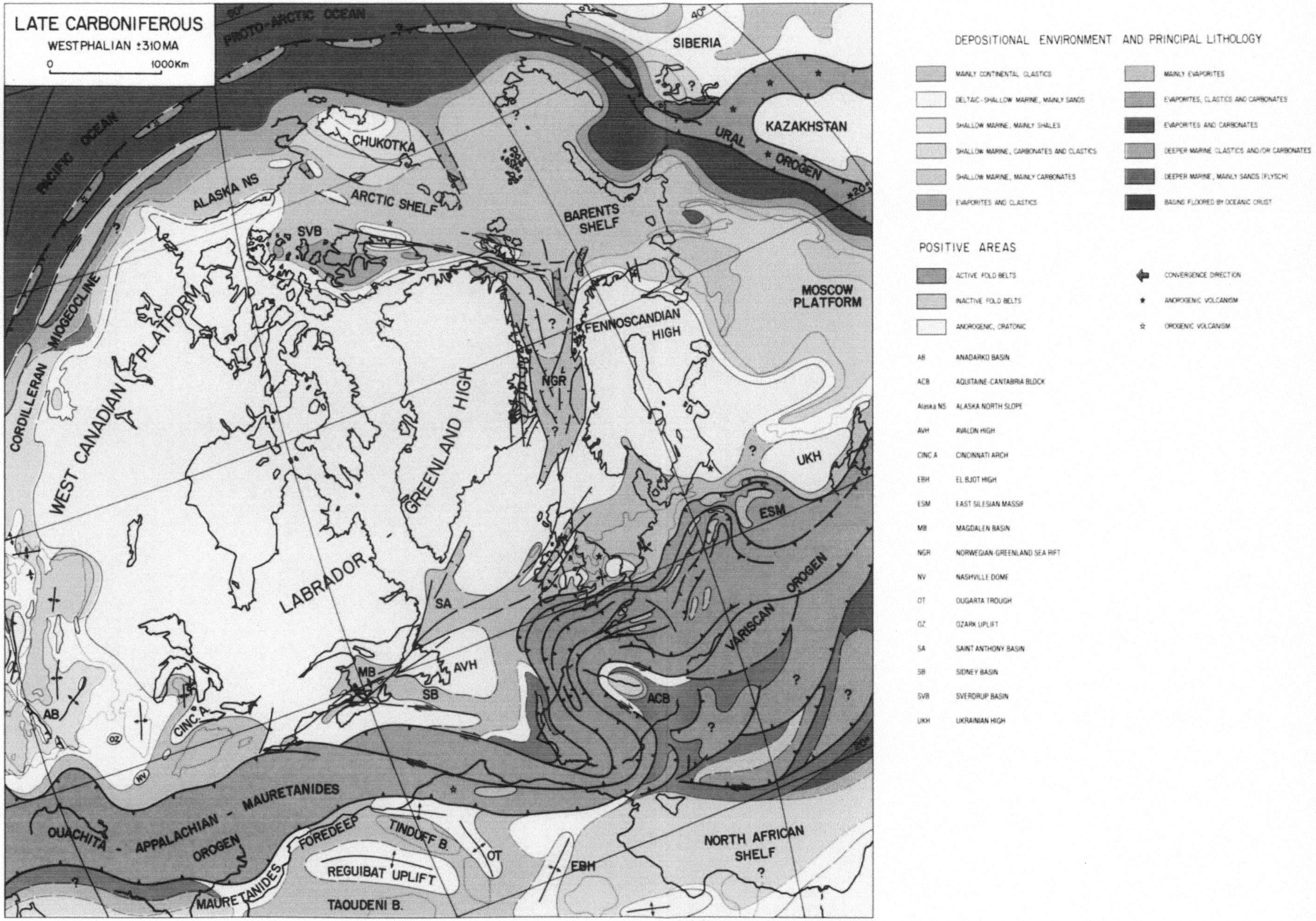

DEPOSITIONAL ENVIRONMENT AND PRINCIPAL LITHOLOGY

MAINLY CONTINENTAL CLASTICS

DELTAIC: SHALLOW MARINE, MAINLY SANDS

SHALLOW MARINE, MARLY SHALES

SHALLOW MARINE, CARBONATES AND CLASTICS

SHALLOW MARINE, MAINLY CARBONATES

EVAPORITES AND CLASTICS

MAINLY EVAPORITES

EVAPORITES, CLASTICS AND CARBONATES

EVAPORITES AND CARBONATES

DEEPER MARINE CLASTICS AND/OR CARBONATES

DEEPER MARINE, MAINLY SANDS (FLYSCH)

BASINS FLOORED BY OCEANIC CRUST

→ CONVERGENCE DIRECTION

★ ANOROGENIC VOLCANISM

☆ OROGENIC VOLCANISM

POSITIVE AREAS

ACTIVE FOLD BELTS

INACTIVE FOLD BELTS

ANOROGENIC, CRATONIC

AB   ANADARKO BASIN
ACB   AQUITAINE CANTABRIA BLOCK
Alaska NS ALASKA NORTH SLOPE
AVH   AVALON HIGH
CINC A CINCINNATI ARCH
EBH   EL BJOT HIGH
ESM   EAST SILESIAN MASSIF
MB   MAGDALEN BASIN
NGR   NORWEGIAN GREENLAND SEA RIFT
NV   NASHVILLE DOME
OT   OUGARTA TROUGH
OZ   OZARK UPLIFT
SA   SAINT ANTHONY BASIN
SB   SIDNEY BASIN
SVB   SVERDRUP BASIN
UKH   UKRAINIAN HIGH

LATE CARBONIFEROUS
WESTPHALIAN ±310 MA

0   1000 Km

SIBERIA

KAZAKHSTAN

URAL OROGEN

MOSCOW PLATFORM

BARENTS SHELF

FENNOSCANDIAN HIGH

UKH

ESM

VARISCAN OROGEN

NORTH AFRICAN SHELF

ACB

EBH

OT

PROTO-ARCTIC OCEAN

CHUKOTKA

ARCTIC SHELF

ALASKA NS

GREENLAND HIGH

NGR

SVB

LABRADOR

SA

MB

AVH

SB

REGUIBAT UPLIFT

TINDUFF B.

FOREDEEP

TAOUDENI B.

MAURETANIDES

MAURE...

PACIFIC OCEAN

CORDILLERAN MIOGEOCLINE

WEST CANADIAN PLATFORM

OUACHITA - APPALACHIAN - APPALACHIAN OROGEN

CINC A

NV

AB

Plate 10

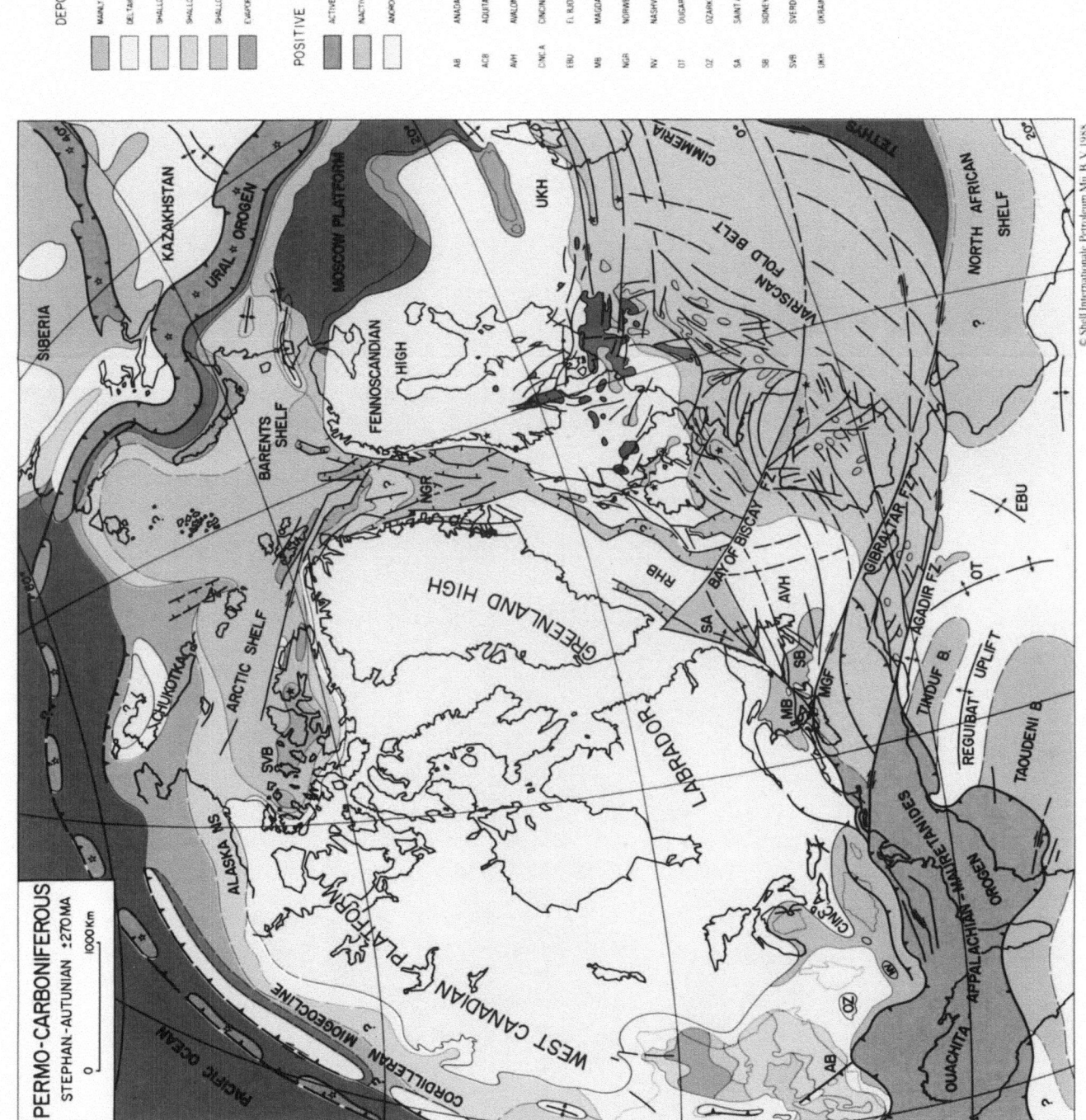

PERMO-CARBONIFEROUS
STEPHAN-AUTUNIAN ±270MA

0   1000Km

DEPOSITIONAL ENVIRONMENT

- MAINLY CONTINENTAL CLASTICS
- DELTAIC, SHALLOW MARINE, MAINLY SANDS
- SHALLOW MARINE, MAINLY SHALES
- SHALLOW MARINE, CARBONATES AND CLASTICS
- SHALLOW MARINE, MAINLY CARBONATES
- EVAPORITES AND CLASTICS

AND PRINCIPAL LITHOLOGY

- MAINLY EVAPORITES
- EVAPORITES, CLASTICS AND CARBONATES
- EVAPORITES AND CARBONATES
- DEEPER MARINE CLASTICS AND/OR CARBONATES
- DEEPER MARINE, MAINLY SANDS (FLYSCH)
- BASINS FLOORED BY OCEANIC CRUST

→ CONVERGENCE DIRECTION
★ ANOROGENIC VOLCANISM
✷ OROGENIC VOLCANISM
▮ PLATEAU BASALTS

POSITIVE AREAS

- ACTIVE FOLD BELTS
- INACTIVE FOLD BELTS
- ANOROGENIC, CRATONIC

- AB   ANADARKO BASIN
- ACB   AQUITAINE CANTABRIAN BLOCK
- AVH   AVALON HIGH
- CINC A CINCINNATI ARCH
- EBU   EL BUET UPLIFT
- MB   MAGDALEN BASIN
- NGR   NORWEGIAN GREENLAND AND SEA RIFT
- NV   NASHVILLE DOME
- OT   OUGARTA TROUGH
- OZ   OZARK UPLIFT
- SA   SAINT ANTHONY BASIN
- SB   SIDNEY BASIN
- SVB   SVERDRUP BASIN
- UKH   UKRAINIAN HIGH

Plate 11

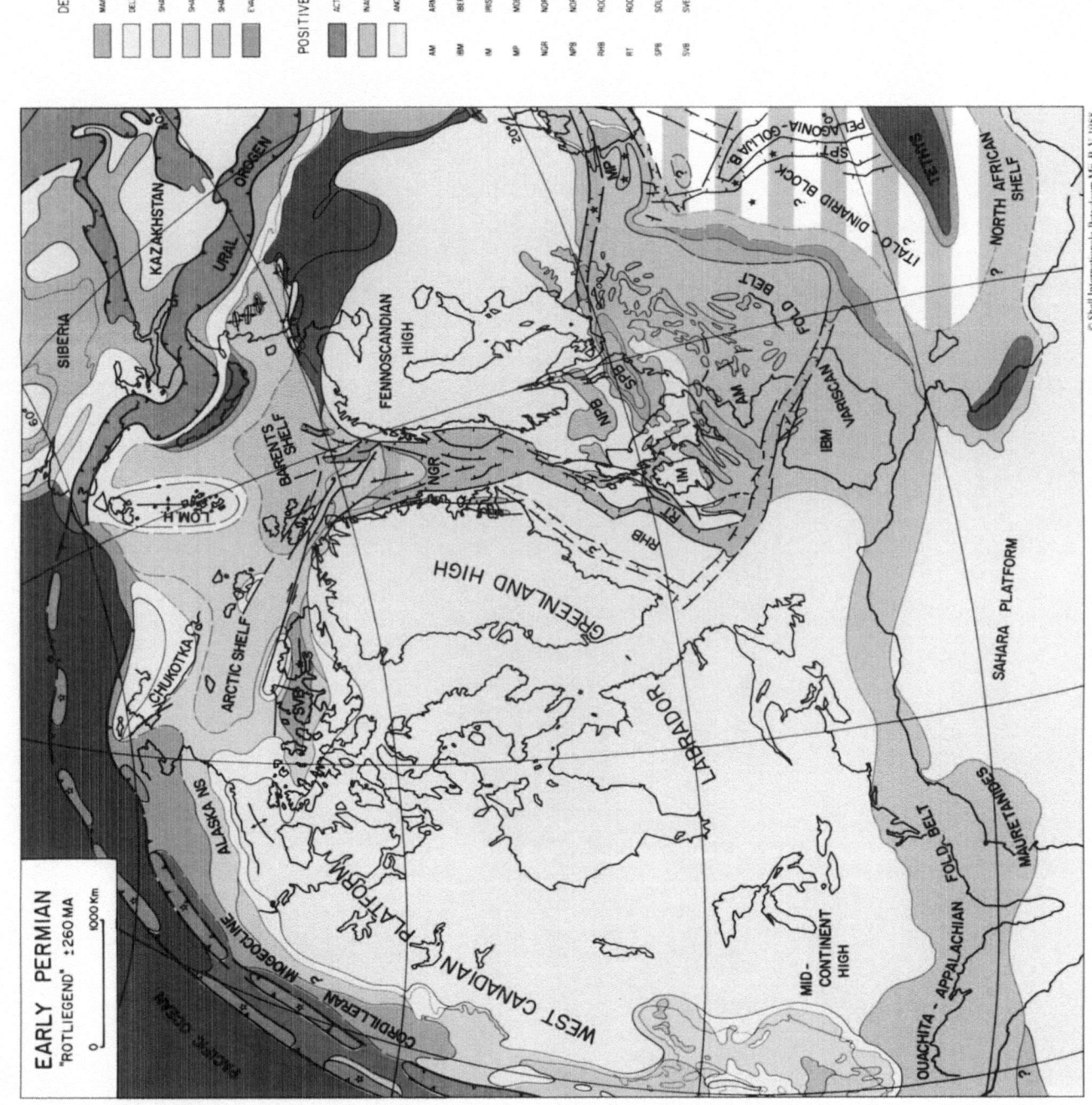

EARLY PERMIAN
"ROTLIEGEND" ±260 MA

0   1000 Km

DEPOSITIONAL ENVIRONMENT AND PRINCIPAL LITHOLOGY

MARLY CONTINENTAL CLASTICS
DELTAIC, SHALLOW MARINE, MAINLY SANDS
SHALLOW MARINE, MAINLY SHALES
SHALLOW MARINE, CARBONATES AND CLASTICS
SHALLOW MARINE, MAINLY CARBONATES
EVAPORITES AND CLASTICS

MAINLY EVAPORITES
EVAPORITES, CLASTICS AND CARBONATES
EVAPORITES AND CARBONATES
DEEPER MARINE, CLASTICS AND/OR CARBONATES
DEEPER MARINE, MAINLY SANDS (FLYSCH)
BASINS FLOORED BY OCEANIC CRUST

CONVERGENCE DIRECTION
ANOROGENIC VOLCANISM
OROGENIC VOLCANISM

POSITIVE AREAS

ACTIVE FOLD BELTS
INACTIVE FOLD BELTS
ANOROGENIC, CRATONIC

AM   ARMORICAN MASSIF
IBM   IBERIA MESETA
M   HIGH MASSIF
MP   MOESIAN PLATFORM
NGR   NORWEGIAN-GREENLAND SEA RIFT
NPB   NORTHERN PERMIAN BASIN
RHB   ROCKALL-HATTON BANK
RT   ROCKALL TROUGH
SPB   SOUTHERN PERMIAN BASIN
SVB   SVERDRUP BASIN

© Shell Internationale Petroleum Mij. B.V. 1988

Plate 12

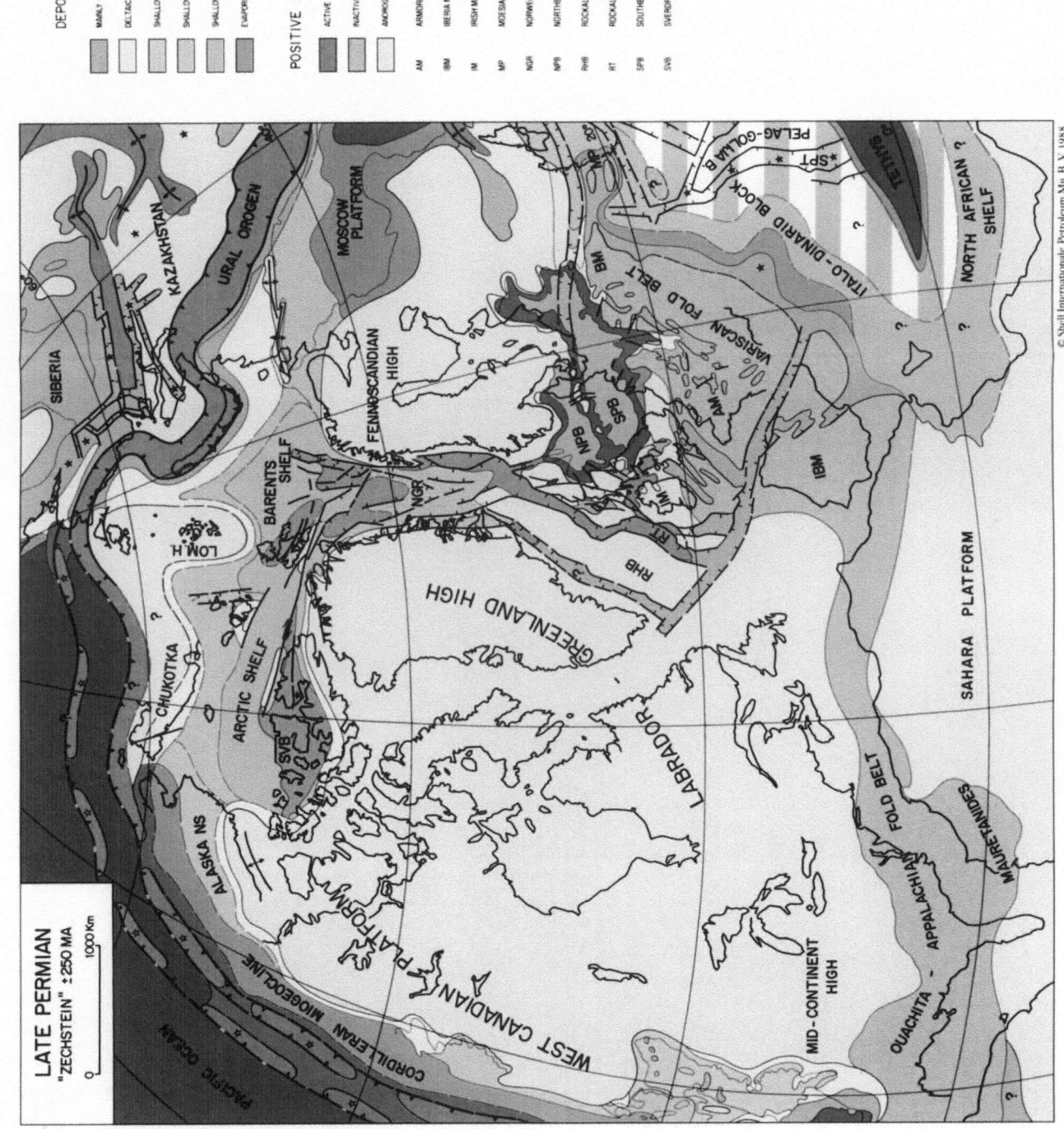

DEPOSITIONAL ENVIRONMENT AND PRINCIPAL LITHOLOGY

- MAINLY CONTINENTAL CLASTICS
- DELTAIC-SHALLOW MARINE, MAINLY SANDS
- SHALLOW MARINE, MARLY SHALES
- SHALLOW MARINE, CARBONATES AND CLASTICS
- SHALLOW MARINE, MAINLY CARBONATES
- EVAPORITES AND CLASTICS

- MAINLY EVAPORITES
- EVAPORITES, CLASTICS AND CARBONATES
- EVAPORITES AND CARBONATES
- DEEPER MARINE, CLASTICS AND/OR CARBONATES
- DEEPER MARINE, MAINLY SANDS (FLYSCH)
- BASINS FLOORED BY OCEANIC CRUST

→ CONVERGENCE DIRECTION
★ ANOROGENIC VOLCANISM
☆ OROGENIC VOLCANISM

POSITIVE AREAS

- ACTIVE FOLD BELTS
- INACTIVE FOLD BELTS
- ANOROGENIC, CRATONIC

AM   ARMORICAN MASSIF
IBM IBERIA MESETA
IM   IRISH MASSIF
MP   MOESIAN PLATFORM
NGR NORWEGIAN-GREENLAND SEA RIFT
NPB NORTHERN PERMIAN BASIN
RHB ROCKALL-HATTON BANK
RT   ROCKALL TROUGH
SPB SOUTHERN PERMIAN BASIN
SVB SVERDRUP BASIN

LATE PERMIAN
"ZECHSTEIN" ±250 MA

0   1000 Km

PACIFIC OCEAN

CORDILLERAN MIOGEOCLINE

ALASKA NS

WEST CANADIAN PLATFORM

MID-CONTINENT HIGH

OUACHITA - APPALACHIAN FOLD BELT

MAURETANIDES

SAHARA PLATFORM

LABRADOR

GREENLAND HIGH

ARCTIC SHELF
SVB

CHUKOTKA

L.OM.H.

SIBERIA

KAZAKHSTAN

URAL OROGEN

MOSCOW PLATFORM

BARENTS SHELF

FENNOSCANDIAN HIGH

NGR

NPB

SPB

BM

IM

RT

RHB

IBM

AM

VARISCAN FOLD BELT

ITALO-DINARID BLOCK

PELAG-GOLLA E.

SPT

TETHYS

NORTH AFRICAN SHELF

© Shell Internationale Petroleum Mij B.V. 1988

Plate 13

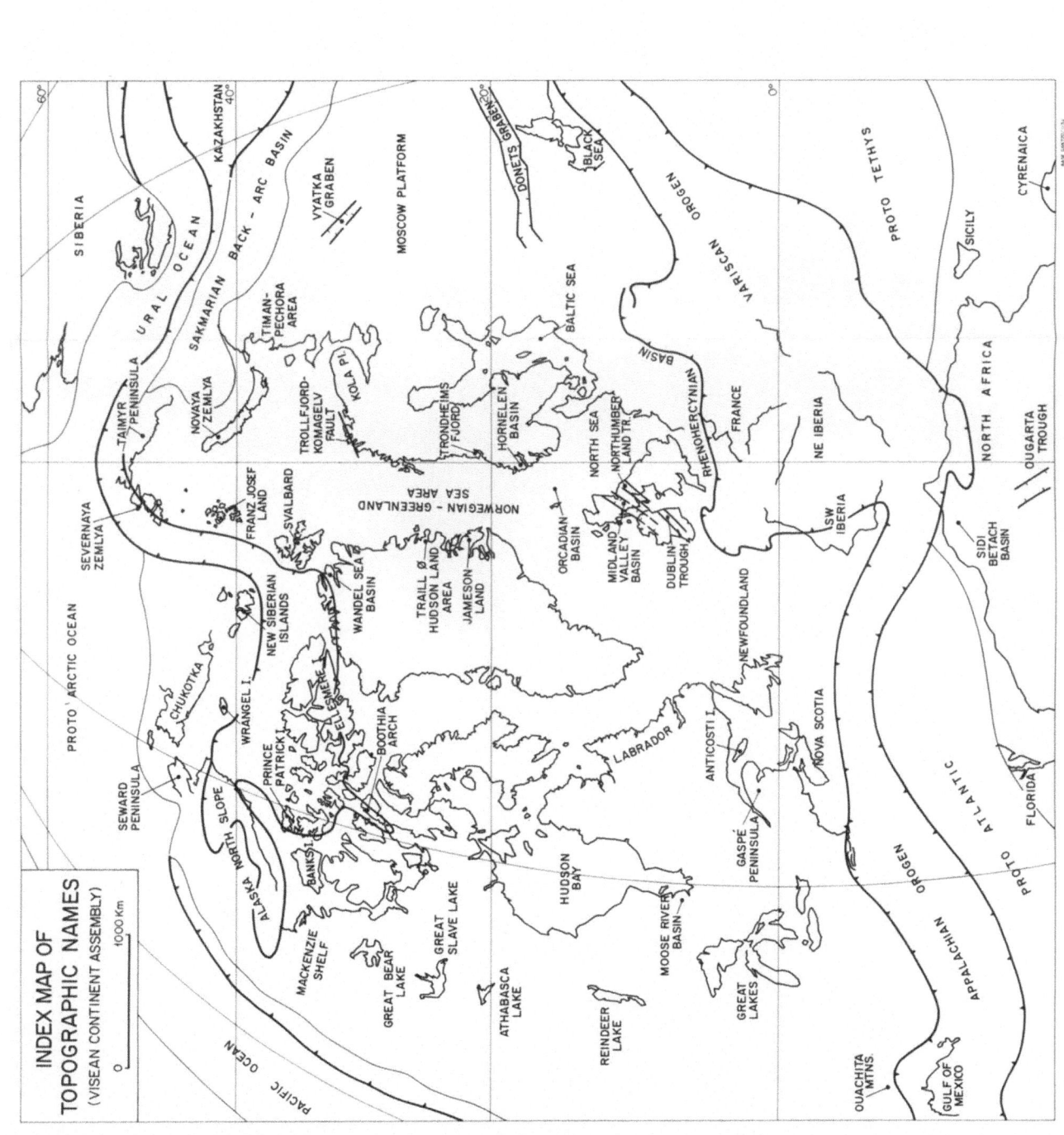

INDEX MAP OF
TOPOGRAPHIC NAMES
(VISEAN CONTINENT ASSEMBLY)

0   1000 Km

Plate 14

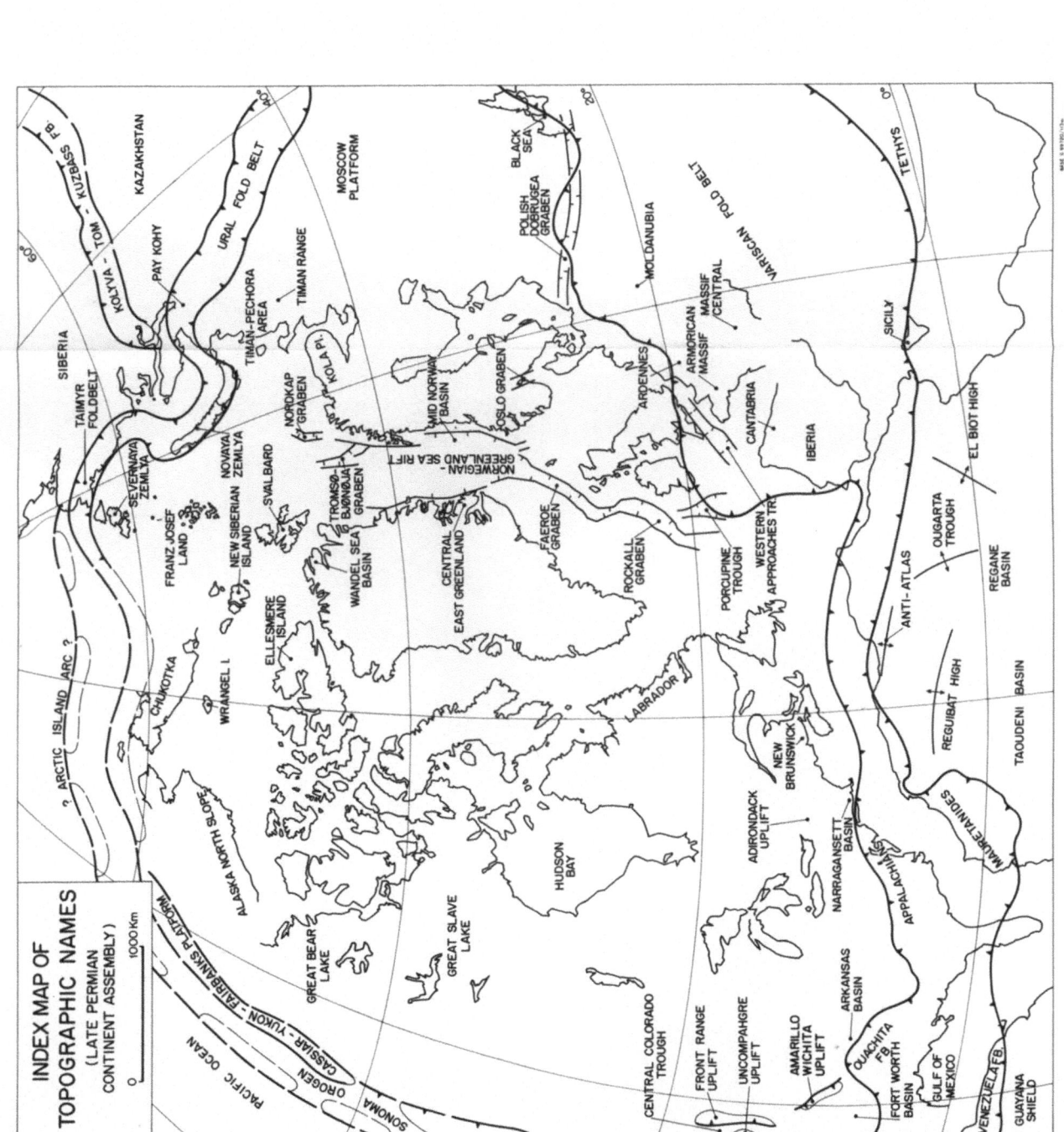

INDEX MAP OF
TOPOGRAPHIC NAMES
(LATE PERMIAN
CONTINENT ASSEMBLY)

0   1000 Km